HUBBARD
THE FORGOTTEN BOEING AVIATOR

JIM BROWN

Seattle, Washington
Portland, Oregon
Denver, Colorado
Vancouver, B.C.

ISBN # 0-89716-651-5
LOC 96-069416
11.0076
Cover design by David Marty
Editing & Production: Elizabeth Lake

Second printing November 1996
10 9 8 7 6 5 4 3 2

Peanut Butter Publishing
226 2nd Avenue West • Seattle, WA 98119
Old Post Office Bldg. • 510 S.W. 3rd • Portland, OR 97201
Cherry Creek • 50 S. Steele • Suite 850 • Denver, CO 80209
Su. 230 1333 Johnston St. Pier 32 Granville Isl., Vancouver, B.C. V6H 3R9
e mail: P NUT PUB@aol.com
http://www.pbpublishing.com

Printed in Canada

Table of Contents

Photo Credits 5
Acknowledgments 7
Foreword..... 9
Introduction 11
Chapter 1
The Early Flying Days 15
Chapter 2
North America's First International Air Mail 31
Chapter 3
Post-War Years 41
Chapter 4
First Air Taxi 43
Chapter 5
First Air Mail Victoria - Seattle and Return 49
Chapter 6
Honeymoon Pilot 55
Chapter 7
First Foreign Aircraft to Visit Victoria 59
Chapter 8
Eddie Rickenbacker Compliments Eddie Hubbard 61
Chapter 9
The Popular Aerial Taxi 63
Chapter 10
San Francisco Aeronautical Show 83
Chapter 11
Seattle's First Air Field 87
Chapter 12
Canadian Crash Inquiry 91
Chapter 13
United States First Foreign Air Mail Contract 95
Chapter 14
Interesting Passengers 129
Chapter 15
Hunting Accident 137

Chapter 16
Bank Robbers and Bloodhounds 139
Chapter 17
Victoria B-1 Crash 141
Chapter 18
Demonstrating Navy Trainers 145
Chapter 19
San Francisco-Chicago Mail Contract 149
Chapter 20
Tributes to a Great Aviator and Airline Builder 171
Chapter 21
Hubbard Was Predecessor of United Air Lines 181
Chronology 184
THE B-1 191
CL-4S 199
MODEL 40 201
MODEL 80 207
Bibliography 210
Index 215
About the Author 229
Order Form 231

Photo Credits

No. 1 - Montana Historical Society

Nos. 2, 4, 5, 6, 7, 8, 9,10,12,17, 20, 26, 27, 28, 29, 33, 37, 43, 46, 47, 48, 49, 50, 51, 62, 53, 54, 57, 58, 59, 60, 61, 62, 63, 63, 66, 68 - The Boeing Company

No. 3 - City of Vancouver B.C. Archives

No. 11 - Museum of History & Industry, Seattle

No. 14 - The Daughters of the Pioneers of Washington, Whidbey Island Chapter #6

Nos. 15, 18, 23 - British Columbia Provincial Archives

Nos. 16, 21, 30, 31, 38, 39 - Pemco Webster & Stevens Collection Museum of History & Industry, Seattle

Nos. 19, 26, 32, - Museum of Flight Collection - Suzanne Graham de Carvalho-Maia

No. 22 - Peter M. Bowers Collection

No. 24 - Campbell River Museum and Archives Photo #796 Alan Greene Collection

No. 34 - Mary Weiler

No. 35 - Jane Galbraith O'Neill

No. 36 - Gordon Green

Nos. 40, 41, 42 - Gail Crosby

No. 44 - Suzanne Graham de Carvalho-Maia Collection

Nos. 45, 56, 66, 66 - Jim Brown

Nos. 66, 69 - Chevron

No. 67 - Sheldon Luck Collection

No. 70 - Museum of Flight

Insert Photos

Photos A, D, G, L - Pemco Webster & Stevens Collection, Museum of History and Industry, Seattle.

Photos B, E, F, K, M, R - The Boeing Company

Photo C - Island County Historical Society, Whidbey Island

Photos I, J - Museum of History and Industry, Seattle

Photo N - Gail Crosby

Photo O - Jim Brown

Photo P - Suzanne Graham de Carvalho-Maia Collection

Photo Q - Gordon Green

Acknowledgments

First I wish to thank my wife, golfing and bridge partner, Gwen, whose encouragement and help made this book a reality.

To Marilyn Phipps and Tom Lubbesmeyer of The Boeing Company Historical Archives for their many hours of patiently finding material and photographs.

Dean Thornton, retired President of Boeing Commercial Airplane group and executive Vice President of the Boeing Company who suggested Brien Wygle to do the foreword. Brien Wygle, retired Vice President of Flight Operations, Boeing Company, whom I thank for reading the manuscript, liking what he read, then doing such a fine job on the foreword.

To Fred and Margaret Brewis, wonderful friends since university days, who gave me a bed and great food while spending many hours of research in Seattle. Fred arranged appointments and guided me along the path to publishing. That is what true friendship is all about.

To Lorne and Peggy Tomalty, also great friends from college days, who gave me a bed and breakfasts when doing research in Victoria.

I also wish to thank the following individuals and organizations: Eddie Hubbard's granddaughter, Suzanne Graham Carvalho-Maia; Ian Baird; Library - University of Washington; Library - University of Victoria; Victoria City Library; Museum of History & Industry, Seattle; Museum of Flight, Seattle; Vancouver City Archives, British Columbia Provincial Archives; City of Victoria Archives; Campbell River Museum & Archives; Marjorie Lee Mortensen, Nevada Historical Society; Sue Karren, United States Archives, Seattle; Mike Pavone,

Seattle; Robin Clarke, Victoria; Van Peirson, Seattle; Cary Goulson, Victoria; Cec Ridout, Victoria; U.S. National Archives & Records Administration; Office of the Postmaster General, National Archives of Canada; Ted Cressy; Patrick & Eva Cutts for the generous use of their copy machine; Office of the U.S. Postmaster General; U.S. National Archives & Records Administration; U.S. Postal Service Library reference department; Diplomatic Branch United States National Archives; National Archives of Canada; William M. Leary; Mary Weiler; Harry Barnes Jr.; George Benson; Dick Malott; Brud Nute; Jack Schofield; Canadian Postal Archives; Jefferson County Historical Society; American Philatelic Research Library; Salt Lake Tribune Library; Montana Historical Society; Greg Brewis; Rick Caldwell; Cathy White; Gail Crosby; Jim Wilson; Harriet U. Fish; James A. Morrow; Elwood White; Gordon Green.

Foreword

The early days of aviation have again captured our imagination after having drifted out of memory for decades. In the last twenty years true stories of fragile machines and bold pilots have invaded the domains of magazines, books and films; aviation museums have risen to prominence and at local airports one may often see resurrected versions of once famous flying machines. The biographies of the founders of airlines and aircraft companies have reminded us of how these huge corporations grew from tenuous beginnings. Jim Brown has chosen an individual from those early times and told his story.

Eddie Hubbard was closely connected with William Boeing and between them they gave birth to the Air Mail. Hubbard has left us no legacy like that giant industrialist, but this book reveals how much he contributed to the beginnings of the industry that has now become Commercial Aviation. History buffs know how Boeing stuck to his dream until by 1934 his vast assembly of aircraft manufacturers and airlines came to dominate the industry. The spark for all this was started by Eddie Hubbard.

The shores of Lake Union and the Duwamish River provide the backdrop for the book, and the cities of Seattle, Vancouver and Victoria provide the setting for North America's first international air mail. This part of the continent is not known for good year-round flying weather and the difficulties of navigating those flimsy machines is wonderfully described. The Air Mail was truly the beginning of commercial aviation and the benefits of faster mail service spurred the development of aircraft that could do the job better. The book describes the development of Boeing airplanes to meet the Air Mail needs and how Hubbard persuaded Bill Boeing to bid for an air mail contract at a rate far below his competitors. Despite the dire predictions of others. Boeing made a substantial profit. He and his successors have led the industry to the wonderful airliners of today that allow the masses to travel the world in comfort, speed and safety.

Jim Brown calls himself an "aerophilatelist" and his book returns time and again to the Air Mail. Not only has he excellent credentials as a collector, but his research in writing this book reveals him as a true historian.

Brien Wygle

Brien Wygle joined the Boeing Airplane Company in 1951 as an engineering test pilot. His career spanned 39 years as Director of Flight Test. He retired in 1990 as Vice President of Flight Operations. As an active test pilot he flew Boeing airplanes from the B-47 to the 757 and 767. He also performed the first flight of the 737 and was co-pilot on the first 747 flight.

Introduction

In the 1920's Eddie Hubbard's name was a household word in the Pacific Northwest and the southern part of British Columbia. He was truly an international man and what could be more appropriate than Eddie piloting, along with William E. (Bill) Boeing, the first North American International air mail from Vancouver, British Columbia to Seattle, Washington, March 3,1919.

In the short period of his aviation career this man received more press than many prominent politicians. He was always at the front whether it be flying, instructing, delivering air mail, running an Aerial Taxi business, as a test pilot or an airline builder. In this book I have tried to give that press cover flavor. Space has only allowed a few of his many headline-producing events.

He was not only an employee of the Boeing Airplane Company, he was a very special and close friend of his boss Bill Boeing. His input into the Boeing Company has never before been fully recognized. It was Eddie Hubbard who convinced Bill Boeing to bid on the San Francisco to Chicago air mail contract. It was Eddie Hubbard who suggested they resurrect a prototype air mail plane, with an improved metal body, and build 25 for the mail route. This was the first commercial mass produced airplane the company had ever built. It was Eddie Hubbard whose name, along with Bill Boeing's, was on the contract with the United States Post Office for this new air mail route. It was Eddie Hubbard, as Vice President of operations in the new company, Boeing Air Transport, who made money for its first month of operation.

It is Eddie Hubbard who stands with giants of Northwest aviation such as William E. Boeing, Phil Johnson and many others as a Pathfinder Award member from the Museum of Flight.

It is Eddie Hubbard whom Eddie Rickenbacker, America's outstanding WW II air ace, called the best aviator in America. As a collector of air mail covers (envelopes) researching the life of an air mail pilot has been an interesting venture. I guess the greatest satisfaction in research is uncovering events either unknown or forgotten in history. There were three discoveries that are of interest. The first was the crash on Eddie's air mail route from Seattle to Victoria by Gerald Smith. Eddie was unable to make the run one day and had Smith go to Victoria to pick up the mail. On leaving Victoria, Smith along with the mail bag crashed. This crash was unknown to the American Air Mail Society. The second incident was the forced landing of a Curtiss Jenny Canuck at Coupeville, Whidbey Island, on May 18, 1919, carrying letters from Victoria to Seattle. This was the first air mail between the two cities and a very historic event. I presented this information to the Island County Historical Society, Coupeville, Washington. The event had been forgotten and can now be acknowledged in the new Museum. Murrieal Short, the museum's archivist, found a photograph of the plane which is reproduced in the book. The third area of discovery was being able to prove to and satisfy the American Air Mail Society that certain envelopes were flown by Eddie from Seattle to Victoria and Victoria to Seattle and did not, as had been previously believed, travel by ship.

There is really a fourth item in research and that is meeting so many interesting people along the way who contributed stories, material and photographs which breathe a personality into a book.

I wish to give special thanks to those who have given me the encouragement to keep going and bring many years of a labor of love to its final conclusion. One of the main actors who has been the driving force behind making this bit of Pacific Northwest and Southern British Columbia aviation history possible is: Elliott Wolf, the man behind Peanut Butter Publishing, who liked my story from the first moment we met and needled me to get to the word "fini." Anyone who creates a company with the name "Peanut Butter" can't be all bad!

Elliott gave me an editor, Liz Lake, whose father Ernest Code, was a test pilot for the Boeing Company. Ironically the man she edited a book on was Boeing's second test pilot. Sitting down with a professional editor is quite an experience. Along with many helpful suggestions Liz has done a super job and I am forever indebted.

To Brien Wygle who flattered me in the foreword - thanks a million. To Bill Boeing Jr., who read the manuscript, liked it and gave me encouragement to publish.

Last but not least to Charles Moore whom I now know so well I call him Charlie! By a chance meeting I gave my manuscript to him to read. He then introduced me to Bill Boeing Jr., and finally weakened to the point of saying, "I'll be your producer." So in the final analysis Charlie made it all happen.

To end I would like to point out the amount of international flavor behind many of the participants and this happened long before free trade!

Eddie Hubbard's mother-in-law came from Canada. Brien Wygle was born in Seattle, raised in Alberta, a pilot in the R.C.A.F. Liz Lake's parents came from Victoria and Charlie Moore's father attended private school on Vancouver Island. Who said it's not a small world?

Jim A. Brown
Pender Island, B.C.
June 30,1996

Chapter 1
The Early Flying Days

The story of Eddie Hubbard is a classic example of the American dream, from an orphanage in San Francisco to Boeing Vice President. The following is an explanation of Eddie Hubbard's place in aviation history. First and foremost he considered himself a pilot. His deeds and actions, in so short a life span, put him amongst the world's great men of aviation as an air mail pilot and airline builder.

Many people will be surprised to learn that pioneer Northwest air mail pilot Eddie Hubbard, a close personal friend of Bill Boeing, was the one who suggested Boeing build twenty-five aircraft to carry air mail and passengers from San Francisco to Chicago. Up until that time The Boeing Aircraft Company had only mass produced military aircraft. Hubbard's creative mind convinced Boeing to venture into commercial aircraft manufacturing.

Eddie Hubbard and instructor Terah Maroney at Lake Washington 1916 in Maroney's Curtiss dual control flying boat.

Edward or 'Eddie' Hubbard came to Seattle in 1907 from San Francisco at the age of 18 and in November 1915 became the first graduate of the Aviation School in the Northwest. Eddie had to do prescribed flying tests in a hydroplane over Lake Washington and passed with flying colors. During the tests the machine sideslipped 200 feet. Fortunately he was high enough to level off 100 feet above the lake.

SEATTLE MAN IS GRANTED LICENSE FROM AERO CLUB

Edward Hubbard Is the First Pupil of Aviation School in Northwest.

COMPLETES ALL TESTS.

Eddie started taking flying lessons in August from Terah T. Maroney, at that time a well known instructor in the Seattle area. Ironically Eddie's future employer, Bill Boeing, was detailed by the Aero Club of America to observe the test flights. Unable to be present Boeing nominated Navy pilot G.C. Westervelt to do the job. Later Westervelt worked for Boeing and the two designed the first Boeing-built aircraft, a seaplane designated B & W for Boeing and Westervelt. Only two were manufactured. They were sold to a New Zealand flying school and later used to fly air mail. The B&W is now philatelic history. New Zealand, in 1974, issued a set of four postage stamps commemorating that country's development of air transport and the B & W was included. The B & W was the first aircraft of the Pacific Aero Products Company which became the Boeing Airplane Company, May 9, 1917. A replica built in 1969 hangs in the Museum of Flight in Seattle.

The B & W was the first aircraft manufactured in 1916 by Pacific Aero Products Company. In 1917 the name was changed to the Boeing Airplane Company. Two B & W's were built and later sold to a flying school in New Zealand. One ended its career flying mail.

Boeing pilot Clayton Scott leaving Vancouver, British Columbia March 3,1969 in a replica of the B & W. This was the 50th anniversary of the first North American International air mail flight from Vancouver to Seattle.

This replica was flown to Vancouver, B.C. on March 3, 1969. Veteran pilot Clayton Scott reenacted North America's first international air mail flight from Vancouver to Seattle by Bill Boeing and Eddie Hubbard 50 years prior in Boeing's personal seaplane a modified C-700 which became CL-4S. Upon arriving back in Seattle, Bill Boeing Jr. was on hand to receive the original Canadian Post Office mail bag from Scott at Renton, Lake Washington.

The only changes from the original B & W to the replica were a modern air-cooled engine to replace the Hall-Scott water-cooled engine and a metal instead of a wood propeller.

Bill Boeing Jr.at Renton,Lake Washington March 3,1969 receiving the original Canadian Post Office mail bag flown fifty years previously by his father William E.Boeing and Eddie Hubbard from Vancouver,B.C. to Seattle. Piloting the B & W replica is the Boeing Company chief production test pilot, Clayton Scott. This aircraft now hangs in the Museum of Flight at Boeing Field, Seattle. The original flight March 3, 1919 was made in Boeing's personal seaplane the CL-4S.

In 1916 the *Seattle Post-Intelligencer* reported, "In order to win the license which makes him eligible to enter any competitive flying event in the United States and other countries, Hubbard was required to make five complete figure eights in the air between two posts set 500 yards apart, and descend without power. He then had to make five more figure eights and bring the airplane to a stop within 164 feet of a given mark. The third requirement was to climb to an altitude of 1000 feet and glide to the surface of the lake. Mr. Hubbard fulfilled these requirements and on the last went 1,500 feet into the air and shut off his engine, making a good landing. Aviator Maroney and Eddie Hubbard are now the only men in the Northwest holding licenses from the Aero Club of America."

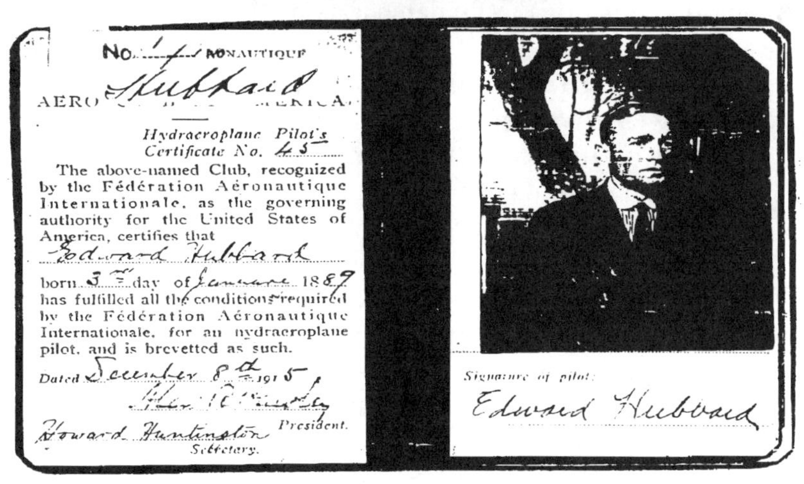

Hydroaeroplane Pilot's Certificate No. 45

The above-named Club, recognized by the Fédération Aéronautique Internationale, as the governing authority for the United States of America, certifies that Edward Hubbard born 3 day of January 1889 has fulfilled all the conditions required by the Fédération Aéronautique Internationale, for an hydroaeroplane pilot, and is brevetted as such.

Dated December 8th 1915

President.

Howard Huntington
Secretary.

Signature of pilot: Edward Hubbard

June 1916 Eddie visited Salt Lake City and the *Tribune* published his outstanding article on the theory of flight and visions of aviations future headed:

AEROPLANE IMPORTANT FACTOR
EXPERIENCED AVIATOR NEEDED

Eddie ably describes the evolution of flight, giving a great deal of credit to Samuel P. Langley a professor at the Smithsonian Institute who in May, 1896 flew a self-propelled, self-controlled model airplane for three-quarters of a mile on the Potomac River. Had Langley not been the subject of criticism by people who considered him unbalanced, we no doubt would have seen men fly long before the Wright Brothers. The United States Post Office honored Langley by issuing a stamp in 1988 commemorating his contributions to aviation. Hubbard went on to praise John Montgomery, who demonstrated with a glider dropped from a balloon at 4,000 feet, that perfect control of weight supporting flat surfaces was possible. He gives a great deal of credit to the Wright Brothers. Then came the ingenious mind of Glenn H. Curtiss whose ability to make light gasoline motors put the United States at the top of the list by winning the first Gordon Bennett aviation cup in 1908 at Rheims, France. With a very limited formal education Hubbard's grasp of the fundamentals of flight were amazing. He went on to say,

> "The aeroplane of today is a highly efficient power-driven vehicle, comprising the most complete unit of a mechanically driven moving body. It is self-contained and can travel on land or water and in the air. It is the fastest mode of transportation controlled by man. This wonderful machine <u>that is to play such an important part in the future of man</u>, is the most abused article of the present day. Men who are supposedly intelligent totally ignore it, while others are attracted by the machine as a spectacular object, and most men consider the operator a fool who holds his life lightly.
>
> As there is a certain amount of danger attached to flying, we should honor those whose lives have been given up to perfect the machine which is rapidly becoming more useful and more safe to mankind.
>
> The theory on which the aeroplane is based is that a horizontal plane with a slightly curved surface, driven through the air with a slight angle of inclination, will so displace the atmosphere that the air will be forced

away from the upper surface, thereby creating a partial vacuum immediately above the plane. As the pressure of the air slightly increases on the lower side of the surface it will readily be seen that the machine will be forced upward in the attempt of the air to equalize in pressure. It is for this reason that the aircraft must be kept under motion to offset the force of gravity and sustain itself in midair."

Hubbard's insight into the future was uncanny. His article goes on to say,

"An aeroplane cannot hover over one spot, and the present style of plane will be unable to do so until motors are developed with a power and lightness that will enable the thrust of the propeller to overcome weight. Should this happen, the machine will be able to stand perfectly still over one spot, pointing straight downward. The day of this achievement is not far away.

It was the development of this feature of the aeroplane that was largely responsible for the death of one of the world's most skillful and daring fliers - Lincoln Beachey. Beachey so constructed an aeroplane that the thrust of his propellers very nearly overcame the weight of the machine. In doing so he sacrificed weight and strength, which meant safety. One wing of his high-powered monoplane crumbled in the air and Beachey fell 500 feet into the water and drowned."

Eddie went on to discuss stalling in great detail. It was his complete understanding of how to avoid a stall that prevented him from a fatal accident and he estimated 50 percent of all aircraft accidents were due to stalling. At the same time another aviator and inventor on the East Coast, Lawrence Sperry was also mastering the stall.

Part of Hubbard's pitch was the development of aviators and he had this to say,

" Millions have been spent in the development of the aeroplane but nobody seems to pay much attention to the development of the most vital part of the whole game - the aviator. He has been allowed to get what knowledge he can the best way he can. Without his skill the aeroplane would soon become a useless article. Why should we not encourage the art of flying and keep in time with the development of the aeroplane.

The European war has shown the value of trained men, and in times of peace men can be trained for less than half the expense required in time of war. This being true, why not pay to train ten men and buy one machine rather than buy ten machines and train one man."

Aviation in America succeeded because of visionaries like Eddie Hubbard and General 'Billy' Mitchell, the much-maligned advocate of a separate service for the Air Force.

In presenting his thoughts to the *Salt Lake Tribune* Hubbard never realized that eleven years later, he would be Vice President and Operations Manager for Boeing Air Transport Inc., headquartered in Salt Lake City.

The Seattle Times had this headline:

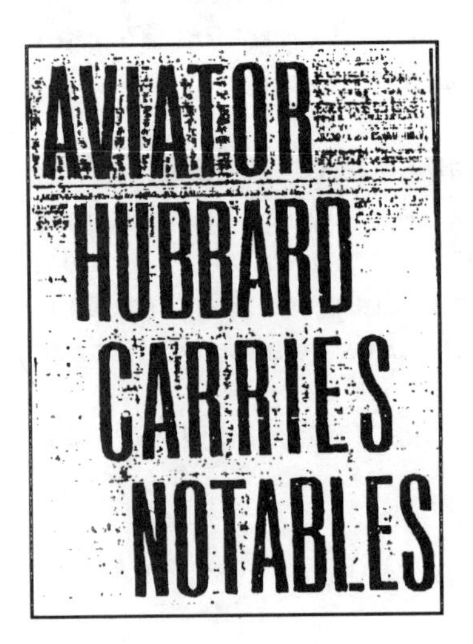
AVIATOR HUBBARD CARRIES NOTABLES

Eddie was not only an outstanding pilot but was a great public relations man. He seemed to sense publicity which landed him time after time on the pages of many newspapers in both the United States and Canada. Hardly a month went by that there wasn't a picture and write up about his exploits.

Mrs. Francis Guy Frink, her daughter, Miss Gloria Frink, of Seattle and Mrs. Betty Umbrecht of Victoria, B.C., were given a ride over Lake Washington in an early flying boat piloted by Eddie. The ladies were attending the Seattle Red Cross Women's Service Camp at Fort Lawton. Mrs. Frink said, " It's just like riding in a big automobile and then rising up in the air just as evenly. Zip and you are up. It is simply wonderful. There isn't anything like it. All my life I have wanted to go up in an airship and now I want to own one!" Mrs. Frink also enthusiastically said on returning to the ground that riding in her automobile was going to seem very tame now. Mrs. Umbrecht went up but Mrs. Frink warned her to put something over her ears as the engines roar made her head ring. She also told Mrs. Umbrecht to say all she had to say before starting the flight because once in the air, although one can talk all she pleases, not a thing can be heard above the noise of the engine.

Bill Boeing got the flying 'bug' and decided to go to Los Angeles and take flying lessons from Glenn Martin. When the course was finished Boeing purchased a ten thousand dollar seaplane from the Martin Company known as a Model TA. It had one pontoon with balancing floats at the outer edge of the lower wings. At the front of the main pontoon was a wheel so the seaplane could ride up a ramp. William Boeing had it crated and shipped home to Seattle. Boeing constructed a canvas hangar on Lake Washington and made his first flight October 21,1915.

Bill Boeing's Martin TA leaving the tent hanger on Lake Washington. His first flight in Seattle was October 21,1915. Boeing purchased the Martin in Los Angeles after learning to fly at the Martin school. A permanent wood hangar later replaced the tent.

Once he had his seaplane in Seattle he began to look for a mechanic. One day, in December 1916, at his friend Arthur Nute's automobile dealership, Nute Motor Company, distributors of the Hudson and Essex, he said to Arthur, "I need a mechanic for my plane."

Arthur replied, "Bill, see the man under that car, he's not only my best mechanic but he also flies an airplane. His name is Eddie Hubbard."

The rest is history. Eddie Hubbard went to work for the Boeing Airplane Company as a test pilot and was later put in charge of Boeing's personal plane. He joined the company's other test pilot, Herb Munter.

Arthur Nute's son Brud observed, "That's how Bill Boeing stole Eddie Hubbard from my father."

Eddie did his first practice flight when he joined Pacific Aero Products Company on February 6,1917 in a B & W. In the company log book he noted, "Found awkwardness in controlling - Turning with wind astern difficult - Front end or body unsteady." At 4:50 p.m. he went up again. "Planed around lake - controls still awkward." The next day he took the B&W up three times and the other test pilot, Herb Munter, was up with the same plane four times that day. Both commented in the log book that controlling was becoming easier.

From February to the end of March, Hubbard and Munter made practice flights in the same aircraft with such log notes as: Speed indicator shows 65 m.p.h. - considerable vibration - new rudder decided improvement - new bracing in body helps to steady front end - ready to make turns just as soon as weather permits - machine took sudden turn with wind and overbanked beyond control landing on left wing tip - landed at new Lake Washington hangar - machine holds turns - motor refused to deliver power due to carb. trouble - recommend strainer be put in gas line - made high speed tests averaging 66.3 m.p.h. - made 2,600 ft. in 16 1/2 minutes - machine gets up on step quickly."

Bill Boeing was a superb chief executive, as noted in his early memorandums. Here is an example of one he sent to his two test pilots, Hubbard and Munter.

MEMORANDUM TO MESSRS. MUNTER AND HUBBARD

February 5th 1917

Commencing today the following instruction governing practice work at the Lake Union Hangar will govern.

Messrs. Munter Hubbard and Locker are detailed to the Hangar and will report every week day at 8 a.m.

As much practice work as possible is to be done by Messrs. Munter and Hubbard, each of whom will have equal call upon the machine.

Every effort is to be made by the men concerned to familiarize themselves with the Dep[1] control. No flights, however, will be made until you have absolutely satisfied yourselves and so reported to me that an instinctive use of the control has been mastered.

If the weather is unpropitious for flying, work will be carried on at the Lake Union shop in connection with the completion of the Dep controls in the machines.

An accurate record of the actual time of running the engine, including the time to warm up, is to be noted in the log in addition to the usual data reported therein.

W E Boeing

(1) Depurdussin - a French aircraft manufacturer. It was a wheel control for the pilot instead of the usual stick control plus an unusual control system for the ailerons.

Eddie left Boeing in August, 1917 to become an instructor at the Rockwell Field Aviation School in San Diego and Herb Munter carried on as the company's only test pilot.

J.C Foley, Boeing Airplane Company Manager, sent two letters to the Navy on Eddie's behalf. They certainly show the high esteem in which the Boeing Company held one Mr. Edward Hubbard.

June 21, 1917

To the Commandant,
Navy Yard, 13th Naval District,
Bremerton, Wash.

Sir:

Please be advised that Edward Hubbard has been employed by the Boeing Airplane Company for the past 6 months as Aviator, and had also had previous experience in this line before entering his present employment.

Mr. Hubbard is a highly capable flyer and has flown several types of machines while in our employ. He is familiar with the Curtiss and Depurdussin types of control.

Mr. Hubbard is a courageous young man, an excellent mechanic and is ably qualified to fill a commission as aviator in the flying corps of the Navy.

Very respectfully yours,
BOEING AIRPLANE COMPANY

JCF - W

August 2, 1917

Chief Signal Officer,
Aviation Section,
San Diego, Cal.
Sir:

The bearer of this letter, Mr. Edward Hubbard, has been in our employ for the six months proceding July 1st. He was employed as Aviator and Mechanic and we found him a very capable young man in both lines of work He is thoroughly familiar with the Curtiss and Dep controls and has flown both Curtiss flying boats and the hydroplane type of machine. He is a sober, industrious and capable young man and ably qualified as an Aviator.

He left our service of his own accord to offer his services to the Government.

Very respectfully,
BOEING AIRPLANE COMPANY

JCF - W

Only commissioned officers were to be used as flight instructors at the various training schools. The services of civilians like Hubbard as of May 1918 were no longer required. Eddie's services, like so many other civilian instructors, were terminated. However he left San Diegans amazed with a show of his aerial skills.

The San Diego Union said this: "It was Hubbard's last flight as a trick flying instructor in the American air service and his wonderful work brought forth a storm of cheers from thousands of blue jackets at Balboa Park as well as expressions of admiration from as many thousands more facinated civilians." Eddie departed San Diego with his usual habit of 'leaving them in the aisles'.

On June 15,1918 Eddie Hubbard rejoined the Boeing Airplane Company and started familiarizing himself with construction details at the plant. Bill Boeing was in Washington, D.C. and instructed Edgar Gott, the company's general manager, in compensation dealings with Hubbard to make some arrangement for reduction of flying hours. He also stated his opposition to storing machines at the plant which eventually led to the hangar on Lake Union.

Gott wired Boeing, "Have closed with Hubbard on the following basis ten dollars per hour, minimum one fifty per month. After thirty hours at above the rate, drops to five dollars per hour to a maximum of fifty hours per month. In other words the maximum salary would be four hundred dollars with no pay for hours above fifty. Time would include in and out of hangar."

Later Eddie approached Gott regarding his salary because flying was his only asset and flyers of equal or lesser ability were obtaining better money. Eddie laid two requests on Gott. His salary be increased to $5,000 per year or $416 a month and he be given a bonus of 25% on any exhibition flights. A mutually satisfactory arrangement was made at a flat $400 a month.

June 20th Eddie started flying once again for the Boeing Airplane Company and made four flights that first day in the C-4. On the first two flights he took Navy personnel on familiarization trips and Edgar Gott had two flips. On the first the motor overheated and on the second a throttle wire broke, forcing a landing.

The Boeing Airplane Company obtained a contract from the Navy to build fifty Model C training seaplanes designated as C-650 to C-699. This was, to the fledgling company, a huge order. One more 'C' class seaplane was built for Bill Boeing's personal use given the designation C-700. This plane was later modified and became known as CL-4S.

During September and October both Hubbard and Berlin, plant superintendent and pilot, made a series of successful flights for the war time Liberty Loan. Hubbard took movie photographer Frank Jacobs up to film a parade for release to movie theatres in Seattle and the West Coast. Gott covered these flights in a letter to Boeing, saying the flights were not as lurid as the newspaper accounts and there was nothing done which would in any way endanger the machine, the pilot, or the people on the street. A few days later Hubbard and Berlin did more Liberty Loan flights with Hubbard doing stunts and Berlin dropping literature and parachutes with flags.

Boeing General Manager Edgar N. Gott sent a memorandum to C.A. Berlin, Supt., E. Hubbard and H.A. Munter: "There is enclosed herewith, for your signature, a form of release from liability while flying, which it is requested that you kindly sign and return to the writer at an early date." A further memorandum on the subject was sent to Berlin, the Supt., also a pilot. "This office has not as yet received release from Mr. Hubbard covering flights. As per conversation with Mr. Hubbard some time ago, the writer understands that he does not wish to sign a formal release, but is not adverse to giving an informal letter to the company stating his position. Kindly see what you can do towards obtaining some record from Mr. Hubbard in this connection." Gott didn't give up and on December 9th he again sent another memorandum to Berlin, "It is noted that Mr. Hubbard has not as yet signed the release from liability while flying. It is requested that you have Hubbard sign this release at earliest possible date and turn form into this office after so doing."

Berlin replied, "The writer has taken this matter up, personally, with Mr. Hubbard and still finds him unwilling to sign the release handed to him some time ago. I went into the matter at some length, and the matter of this company insuring Mr. Hubbard for $410,000 and this company paying the premium was mentioned. Mr. Hubbard spoke favorably of this arrangement, but asked for time to consult Mr. Boeing, personally, in the matter. This, however, was to be done immediately, and this office is to be informed as to the results."

I wonder if Hubbard ever did sign the release form?

Bill Boeing had supreme confidence in his test pilot, as indicated in the following memorandum to Gott headed:

RULES REGARDING FLYING

1. C-700 is not to be flown by anyone excepting Mr Hubbard and myself. Any departure from this regulation is to be at my discretion only. This likewise applies to the carrying of passengers in the machine.
2. Authority is given you to permit such members of the engineering department to ride as passengers in the C-700 for the purpose of conducting tests only. When these tests have been completed, please advise and this authority will be withdrawn.
3. The L-4 motor is to remain in the machine until you receive further advices from me.
4. Mr. Berlin is hereby given authority to continue to fly the C-11 at his own discretion. In event of continued breakages to this equipment, it is my intention to make pilots responsible for damage incurred; however, this order will not be put into effect at present.

W E Boeing

Gott sent a further memorandum to Supt. Berlin:

Two copies of reference are enclosed herewith. It is requested that one of these copies be forwarded to Mr. Hubbard for his information.

In addition to the foregoing please be advised that no women passengers are to be carried in any ship except by written authority from Mr. Boeing. These instructions, of course, pertain only to flying as carried on by the company at present and will not govern at such time as passenger flying on a remunerative basis is embarked upon.

Kindly acknowledge receipt of the above and advise if thoroughly understood.

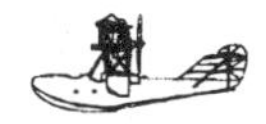

Chapter 2
North America's First International Air Mail

The first experience Eddie had with mail was on July 4, 1916. The Boeing Airplane Company Aviation Log Book has an entry stating Hubbard and Lt. Condon flew a message from the Naval Training Camp at Bremerton to Camp Lewis, which took 52 minutes. Eddie left at 3:10 p.m. and returned with Lt. Munford at 5:10 p.m. from American Lake with an answer to the delivered message. The return trip took 65 minutes.

This was not the first air mail in the State of Washington. On August 10, 1912 Walter Edwards flew mail from Portland, Oregon to Vancouver, Washington. One day later he did it again this time carrying 5,000 pieces of mail. The first in-state air mail was a flight February 20, 1915 in a hydroplane from Tacoma to Seattle by pilot Gustave Stromer.

During February and March 1919, the annual Vancouver, British Columbia Exhibition had a showing of war trophies in the Georgia Street, Horseshow Building. It was for the benefit of the Canadian Red Cross, and a prominent Vancouver druggist, E.S. Knowlton, was a director of the Exhibition and also its chairman. Knowlton must have been part showman and to bring attention to the exhibition he invited W. E. (Bill) Boeing to bring his airplane from Seattle and put on a display of stunt flying over Vancouver. Boeing accepted the invitation and planned to bring his test pilot, Eddie Hubbard, to perform the aerial display.

Refuelling at Anacortes February 17,1919 on first attempt to reach Vancouver, B.C. Damage to the CL-4S dictated returning to Seattle for repairs. Boeing is standing on the pontoon and Hubbard is on the wing.

Boeing and Hubbard left Seattle on February 17th in Boeing's personal seaplane, first classified as C-700. Modifications were made and a new Hall-Scott L4 engine was installed. With this change the aircraft was identified as CL-4S. Departing Lake Union with Bill Boeing at the controls, the plane ran into problems at their first refueling stop in Anacortes. Boeing brought his plane into land at the harbor entrance where large waves were breaking and while he made good on the landing, a sudden gust of wind put the plane up on its tail. Just at that time, a large wave hit them and snapped the rudder controls. Fortunately, a launch that had been standing by towed the disabled craft to a beach. There was no more flying that day and parts were obtained from the Boeing plant at Georgetown, south Seattle. This was the name at that time for the part of Seattle now occupied by the King County airport and Boeing Headquarters. The flight to Vancouver was cancelled.

On February 27th a second attempt was made and again a stop was made at Anacortes for fuel at Standard of California's facilities. Most aircraft owners insisted on Standard's Red

Crown gasoline due to its high octane rating. Charles Lindbergh used Red Crown on his famous trans-Atlantic flight. On this occasion the weather continued to be a problem — it was very windy with snow flurries. After refueling, it was decided to spend the night in Anacortes and attempt to make Vancouver the following day, weather permitting. The morning dawned clear and the two pilots continued their journey to Vancouver, arriving at Coal Harbor, Stanley Park in the afternoon. They secured the seaplane at the Royal Vancouver Yacht Club dock. It is interesting to note that this flight required the approval of the United States Navy Department and was done so at a request of the Canadian Government. Only two persons were allowed to take part, with the official Navy order reading: "Mr. Boeing and Mechanician." Bill Boeing was said to have teased his noted test pilot, "Eddie, you are my *mechanician* today, and don't you forget it!"

Arriving at Vancouver B.C. February 28,1919, landing in Coal Harbor, and tieing up at the Royal Vancouver Yacht Club wharf.

In the skies over the city of Vancouver, Hubbard put on a daring aerial display that entertained thousands of people standing on their porches or on the streets before their homes. While airplanes were no strangers to the Vancouver skies even in 1919, it would have been a very unusual experience to see a seaplane doing aerobatics such as those performed by Eddie Hubbard. On Sunday March 2nd he made two flights, each with a passenger. The first was with Mrs. Jimmie Patterson, wife of a Vancouver lumberman, who was in the air for twenty minutes. Reports said she was the first woman on the North American continent to take part in spinning, tail dives, looping and Immelman turns. One Seattle newspaper called her "A game little society woman who came down after the flight smiling but awfully cold." The second passenger was Lt. J.C. Andrews, an aviator of World War I who credited Hubbard as "the best pilot I've ever seen."

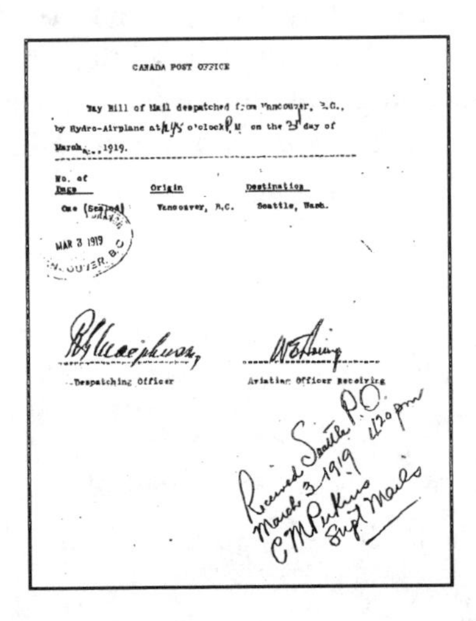
CANADA POST OFFICE

Way Bill of Mail despatched from Vancouver, B.C., by Hydro-Airplane at 1.45 o'clock P.M. on the 3rd day of March, 1919.

No. of Bags	Origin	Destination
One (Sealed)	Vancouver, B.C.	Seattle, Wash.

MAR 3 1919 VANCOUVER, B.C.

R G Macpherson
Despatching Officer

W E Boeing
Aviating Officer Receiving

Received Seattle P.O. 4.20 pm
March 3 1919
C M Perkins
Supt Mails

Canada Post Office waybill for one sealed bag of mail dispatched by Vancouver Postmaster R.(Bob) Macpherson and received by W.E.Boeing who is called an Aviating Officer! The mail was signed for in Seattle by C.M.Perkins, Supt. Mails, United States Post Office.

Exhibition chairman Knowlton, who can be credited with originating this outstanding display, now came up with another idea which would make the event an historic milestone in air mail history. He suggested to his friend, Vancouver Postmaster R.G. (Bob) Macpherson, that Boeing and Hubbard be given a sack of mail to take with them on their return to Seattle. MacPherson thought it was a great idea and provided a new canvas mail bag for the occasion. There was no extra postal charge for the air mail, only the existing surface rate of 3c. Senders were so happy to have a letter flown by the first air mail to Seattle some even got carried away and used two 3c stamps.

Eddie Hubbard standing in the front cockpit and Bill Boeing climbing aboard the CL-4S seaplane with the mail bag. North American air mail history is about to be made.

Vancouver Daily Sun.

With Which Is Incorporated The Daily News Advertiser

No. 02 VANCOUVER, B.C., TUESDAY, MARCH 4, 1919

City Sends First Mail to Seattle by Aeroplane

Lieut. W. E. Boeing With Pilot Eddie Hubbard Left City at 12:58 and Make a Successful Return Flight to Seattle, a Distance of 180 Miles in Two and a Half Hours.

Departure of Birdmen Witnessed by Large Crowd of Citizens Who Kept the Air Craft in Sight Till It Finally Disap...

On March 3rd Knowlton and Macpherson handed Bill Boeing the mail bag for the flight to Seattle via Victoria. For reasons unknown, Victoria was bypassed and the seaplane proceeded to Edmonds, just north of Seattle, to refuel for the trip on to Boeing's hangar at Lake Union. The flight from Vancouver was logged at two hours and ten minutes.

Hubbard and Boeing landing on Lake Union with the first North American International air mail after a fuelling stop at Standard of California's facilities in Edmonds, just north of Seattle.

Hubbard and Boeing at the Lake Union hangar with the sixty air mail letters from Vancouver, B.C. Any that have survived are worth approximately $4,000.

DEPARTMENT OF
PUBLICITY AND INDUSTRIES
402 PENDER ST.
VANCOUVER. CANADA.
VIA AIRPLANE MAIL
First Flight
Vancouver, B.C. to Seattle

Mr. Eddie Hubbard,
c/o Mr. W.E. Boeing,
Seattle, Wash.

Via sea-plane driven by
W.E.Boeing, Vancouver to Seattle.

An air mail collector's dream. One of 60 letters flown. This one is the pride of Eddie's granddaughter Suzanne Graham de Carvalho-Maia.

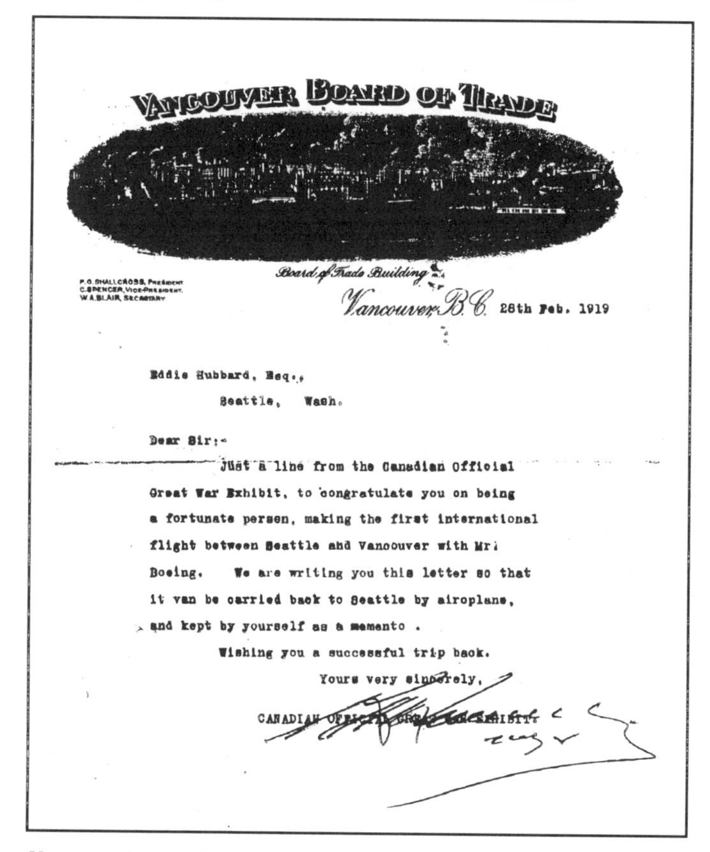
VANCOUVER BOARD OF TRADE

Board of Trade Building

Vancouver, B.C. 28th Feb. 1919

Eddie Hubbard, Esq.,
Seattle, Wash.

Dear Sir:-

Just a line from the Canadian Official Great War Exhibit, to congratulate you on being a fortunate person, making the first international flight between Seattle and Vancouver with Mr. Boeing. We are writing you this letter so that it van be carried back to Seattle by airoplane, and kept by yourself as a memento .

Wishing you a successful trip back.

Yours very sincerely,

CANADIAN OFFICIAL GREAT WAR EXHIBIT.

Vancouver Board of Trade letter to Eddie Hubbard in the above envelope.

History was made! This was the first air mail to cross an international boundary in North America. The mail bag contained sixty letters including one from Postmaster Macpherson to his counterpart in Seattle, Edgar Battle. The prophetic message read,

> "Man made arbitrary lines may show upon the earth's surface, but when we mount up on wings as eagles no line of demarcation then shows between our young Dominion of Canada and her mighty ally to the south, the great United States of America. May the first air mail be the harbinger of thousands more to follow, so that our two countries may know more of each other."
> Macpherson ended this remarkable message with a Robbie Burns quotation,
>
> "Then let us pray that it may be,
> As come it will for all that,
> That man to man this world o'er,
> Shall brithers be and a'that."

Five months after this epic West Coast flight, the Aero Club of Canada flew air mail from Toronto to New York. A private, one dollar commemorative vignette or private stamp was produced for this flight, inscribed: "The First International Aerial Mail Service, August 1919, Toronto - New York." There was still no air mail rate. The covers were flown at the then surface rate of 3c plus the private commemorative $1 issue. It must have come as a shock when the Aero Club found out they were five months late in claiming the first air mail to cross the 49th parallel!

When William Boeing returned to Seattle, he and his Vice President, Edgar Gott, made a presentation to the Seattle Chamber of Commerce and later to Washington D.C., putting forward the view that air mail between Alaska and Seattle was a realistic concept. In his statement, Boeing said, "Large types of seaplanes now available could carry all the mail to and from Alaska and in addition, there would be accommodation for passengers." Boeing's dream would take many years to fulfill.

By the way, if you are fortunate enough to have an envelope or "cover" as collectors call them from this first flight back in March 3, 1919, put it in a safe place — the last one sold by Charles G. Firby Auctions went for a cool $3,850. Several years ago one hit an all-time high of $7,300 in Toronto.

Chapter 3
Post-War Years

Not long after the historic air mail flight from Vancouver, the Boeing Airplane Company arrived at the crossroads of continuing to produce aircraft or close its doors. After producing the 50 Model C trainers for the Navy, the Boeing plant on the Duwamish Waterway in South Seattle received an order from the Government to build fifty Curtiss HS-2L flying boats. With the war over, just about all military aircraft orders were cancelled; however, Boeing got a reprieve — the order for HS-2L's was reduced to 25 or half the original order.

At this period in American aircraft manufacturing, apart from a small plant in San Francisco, the Boeing factory was the only operation of its kind west of Detroit. Their operation was one of the few self-contained aircraft factories in the United States. Everything with the exception of motors, radiators, propellers and instruments was made right at the Duwamish plant.

Bill Boeing commented, "Without further Government aid, it will be a great loss not only to the community but also to the Government if the authorities do not back up an excellent organization, which has taken eighteen months to build. In the event of an emergency we will have to rebuild what is already in place. We are presently engaged in developing types of flying boats adaptable to sporting and commercial use. These boats are actually under construction. The new flying boats are 100 and 200 horsepower with the smaller boat carrying a pilot and two passengers having a flight duration of three hours. The larger one carries a pilot, three passengers and freight with a flight duration of five hours cruising at ninety miles an hour."

What the Seattle airplane designer and manufacturer had in mind for the larger plane was a machine to fulfill his dream of mail and passenger service between Seattle and Alaska.

Once again Eddie became news when he put on a sensational display of flying for the returning men of the 91st Division. As *The Seattle Times* reported, "Most spectacular of any feats in the air which have yet greeted returning war veterans in Seattle were those executed yesterday afternoon over the heads of the parading fighting men of the 91st Division by aviator Edward Hubbard of the Boeing Airplane Company." For more than thirty minutes Eddie, along with passenger Lt. Harry Kay of the Royal Flying Corp. excited relatives and friends of men in the 91st Division who had packed the streets of Seattle. When they returned to the hangar at Lake Union, Lt. Kay told Eddie he didn't believe a seaplane was capable of carrying out the maneuvers he had just been through. Eddie Hubbard was receiving a great deal of attention for his outstanding flying ability.

Chapter 4
First Air Taxi

SENATOR POINDEXTER ABOUT TO START FIRST AIR TRIP

The first major American political figure to take advantage of a relatively new time-saving device, the airplane, was United States Senator Miles Poindexter from the state of Washington. On April 17, 1919, at the invitation of Bill Boeing, Senator Poindexter used the flying skills of Eddie Hubbard to make a visit to the Navy Base at Bremerton. Poindexter was pressed for time to make a trip to Bremerton and Everett in the same day. Upon learning this, Boeing suggested he use an airplane.

The Senator had never flown before but thought it was a great idea. At the Lake Union hangar, Hubbard outfitted Poindexter with sheep-skin lined garments and just before the plane took off Boeing warned his friend to pull the goggles down over his eyes. All Bremerton turned out on the roofs and streets, to say nothing of the men and overall-clad girls employed in the Navy yard. Hubbard thrilled the waiting crowd as he circled over the town, dropping within hailing distance of the spectators as he expertly put the plane onto the water without so much as a splash.

The Boeing Commercial Air Service was formed with Eddie Hubbard as pilot and using the Lake Union hangar at Roanoke Street as the passenger depot.

The first passenger was C.L. Campbell of Seattle who said to Eddie, "I'll go around the world with you. That was the greatest experience I've ever had."

Eddie Hubbard leaving Lake Union May 2,1919 with the Boeing Aerial Taxi's first paying customer, C.L. Campbell.

The next passengers were James Gibbs and Ralph and Alfred Lomen, all Alaska residents, accompanied by Miss Vera Smith of Seattle, Mr. Gibbs fiancee. Alfred was the daredevil of the group and said to Eddie, "If you take me out and loop and tailspin me, I'll pay you five dollars a minute instead of the dollar a minute rate for that mild cross country stuff." The increased revenue must have been a great temptation to the pilot of a fledgling air taxi business but Eddie stuck to his instructions—no aerobatics.

The Lomen brothers were in the reindeer business and talked to Bill Boeing about building a couple of machines for use in Alaska. Boeing Sales Manager Herb Berg also got into the act about the possibilities and practicabilities of an airplane for herding their four big reindeer herds, which numbered 18,000 head in the Nome district. The Lomen brothers pointed out that as much as 400 miles intervened between some of the herds and it took several weeks to get from one herd to another. They shipped 120,000 pounds of reindeer meat to southern markets in the previous year. Many years later Ralph Lomen had this to say about his May 2, 1919 flight with Eddie Hubbard.

"My brother Alfred, James Gibbs and myself were mustered out of the Army at Camp Lewis. While we were in Seattle waiting for the first ship to Nome we went to Lake Union to see the float plane Eddie Hubbard was flying.

"In those days an aviator was a young man's hero. We talked to Eddie and asked if we could hire him to take us up. He told us that since he was flying for the Boeing Airplane Company, Mr. Boeing would have to authorize the flight.

"We promptly called on Boeing in his office at the Hoge Building downtown and learned that the Boeing Company recently had made arrangements for commercial passenger flights. Tickets had been printed in anticipation of the new service starting the next month.

"When Boeing heard our request he told us if Eddie Hubbard is ready, we could go any time convenient to him. The news was relayed to Hubbard and we made an appointment for 2 o'clock the next afternoon. May 2nd was a beautiful sunny day as we made our way to the hangar on Lake Union.

"The first ride that day was complimentary to a member of the press. When our turn came we were given a seagull's view of the City, Lake Washington and the University District. The ride was smooth and we were thrilled beyond words. The 15-minute trip cost us $15 each.

Ralph Lomen, Alfred Lomen, Eddie Hubbard, Miss Vera Smith, James Gibbs and C. L. Campbell in front of the taxi at the Lake Union hangar May 2, 1919. The Aerial Taxi was Bill Boeing's personal seaplane, designated CL-4S.

"After the flight we went to the Boeing plant on the Duwamish Waterway. Phil Johnson, later President of the company, showed us around the factory. His suggestion that each of us purchase $50 worth of stock in the company went unheeded. This investment oversight on my part is the only reason my wife and I haven't taken a trip around the world!"

Apart from the Lomen's party another customer, Mrs. Mildred Layton of Seattle, arrived for a flight. The newest business in Seattle was up and running. The *Seattle P.I.* reported it as a novel way of getting to and from work fast. Herb Berg, Boeing sales manager, was living near Tacoma and it normally took him an hour and a half to drive the distance. When Herb had Eddie take him home from Lake Union in the flying boat, it was all of eight minutes.

Eddie Hubbard was what a good pilot should be—CAREFUL, and he was quick to set the Aerial Taxi rules:

Flights are offered when sightseeing is possible.
Will go anywhere a water landing is available.
All arrangements to be made with the pilot
at the hangar.
Passengers must sign a waiver.

Even employees had to sign a waiver. August 14, 1929 Eddie took his mechanic Les Hubble to Victoria and this is the waiver Les Hubble signed.

No. ______ **Boeing Airplane Company**

Flight permit issued to ______________ {As Passenger / As Pilot}

I hereby agree that the BOEING AIRPLANE COMPANY, or anyone connected in any way with said Company, or both, will not be held liable for any injuries or fatalities whatsoever sustained by me, and expressly waive any claim for such from any cause whatsoever.

For flight in plane No. B-1 *only.* Les Hubble

Date Aug 14 *[Good this day only]*

Approved: to Victoria Flight No. 1066

PRESIDENT
VICE-PRESIDENT
SUPERINTENDENT

Courtesy of Boeing Historical Archives

Another waiver signed by Thurman H. Bane, a Major in the Air Service and approved by VP Edgar Gott:

No. ______ **Boeing Airplane Company**

Flight permit issued to Maj. T. H. Bane {As Passenger / As Pilot}

I hereby agree that the BOEING AIRPLANE COMPANY, or anyone connected in any way with said Company, or both, will not be held liable for any injuries or fatalities whatsoever sustained by me, and expressly waive any claim for such from any cause whatsoever.

For flight in plane No. C-700 *only.* Thurman H Bane

Date 9-22-2[illegible] *[Good this day only]* Major Air Service

Approved: [signature]

PRESIDENT
VICE-PRESIDENT
SUPERINTENDENT

Courtesy of Boeing Historical Archives

When flights were finished Eddie gave each passenger a Certificate of Flight issued by Boeing Airplane Company. Below is certificate No. 360 issued to Juanita Rae and signed by Eddie Hubbard.

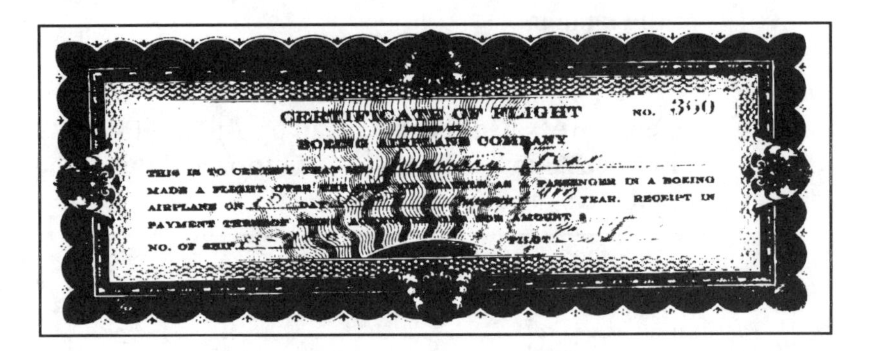

CERTIFICATE OF FLIGHT No. 360

BOEING AIRPLANE COMPANY

THIS IS TO CERTIFY THAT [illegible]
MADE A FLIGHT OVER [illegible] AS A PASSENGER IN A BOEING
AIRPLANE ON [illegible] DAY [illegible] YEAR. RECEIPT IN
PAYMENT THEREOF [illegible] AMOUNT $ [illegible]
NO. OF SHIP [illegible] PILOT [illegible]

Courtesy of Boeing Historical Archives

One month after Eddie starting piloting the Aerial Taxi his fellow Boeing test pilot, Herb Munter, started his own Aerial Taxi business. Munter had left Boeing's employ a week previously and together with his brother started a new venture. They offered an alternative to the Boeing seaplane with a Curtiss Jenny JN-4A Canuck land plane purchased in Canada. They used the municipal golf course at Beacon Hill, now known as Jefferson Park Municipal Golf Course, as a landing field. Later they operated from a field in Kent, just south of town.

Chapter 5
First Air Mail Victoria - Seattle and Return

Two British Flyers Land In Seattle From Victoria

Airmen Bring Invitation Asking Acting Mayor to Attend Victory Pageant in B. C. City on Saturday, May 24.

PATHFINDER MAKES SUCCESSFUL FLIGHT TO SEATTLE; RETURNS IN ONE HOUR, SEVENTEEN MINUTES

The first air mail letters from Victoria, B.C. to Seattle might never have been delivered without help from the United States Army and Eddie Hubbard. Two Royal Flying Corp. pilots from Victoria, B.C., Lt. Robert Rideout and Lt. Harry Brown, in a Curtiss Jenny JN-4A Canuck owned by the Aerial League of Canada,[2] were given the responsibility of delivering a letter to the Mayor of Seattle promoting 24th of May celebrations in Victoria.

Their flight had been delayed by gale-force winds off Port Townsend May 18, 1919. The plane was not making any headway so Rideout decided they must find land and refuel. Sighting a farmer's pea field at Ebey's Prairie, Coupeville, Whidbey Island, Rideout set the plane down. The 'Jenny' was such a curiosity for the people of Whidbey Island they immediately surrounded this intruder from the sky trying to tear pieces of fabric from the aircraft as souvenirs. Fortunately, for Rideout

(2) At the end of WWI the Canadian Government had a large suppy of Curtiss Jenny's built in Toronto. There were many ex-Royal Flying Corp. aviators with nothing to do so the Government formed the Aerial League of Canada and started branches across the country. Each branch was given aircraft to use for barnstorming or any other use to make money. It is interesting to note the Curtiss Company gave only one country permission to make a change in their name and that was the use of the word Canuck for planes made in Canada.

and Brown there was a United States Army base nearby and soldiers offered to guard the plane while the two aviators went for gasoline and oil. Later in the afternoon the wind died down and the Canadian Jenny took off for Seattle before a curious Sunday crowd of over 500 people.

Curtiss Jenny "Pathfinder" forced to land on a farmer's field and refuel at Coupeville, Whidbey Island, Sunday, May 18, 1919. Flight was from Victoria, B. C. to Seattle. Eddie Hubbard met the Canadian Flyers near Seattle and guided them to a landing at Jefferson Municipal Golf course.

THE SEATTLE DAILY TIMES, TUESDAY EVENING, MAY 20, 1919.

CARRY MAIL FROM SEATTLE

TO BRITISH COLUMBIA CAPITAL BY AIR

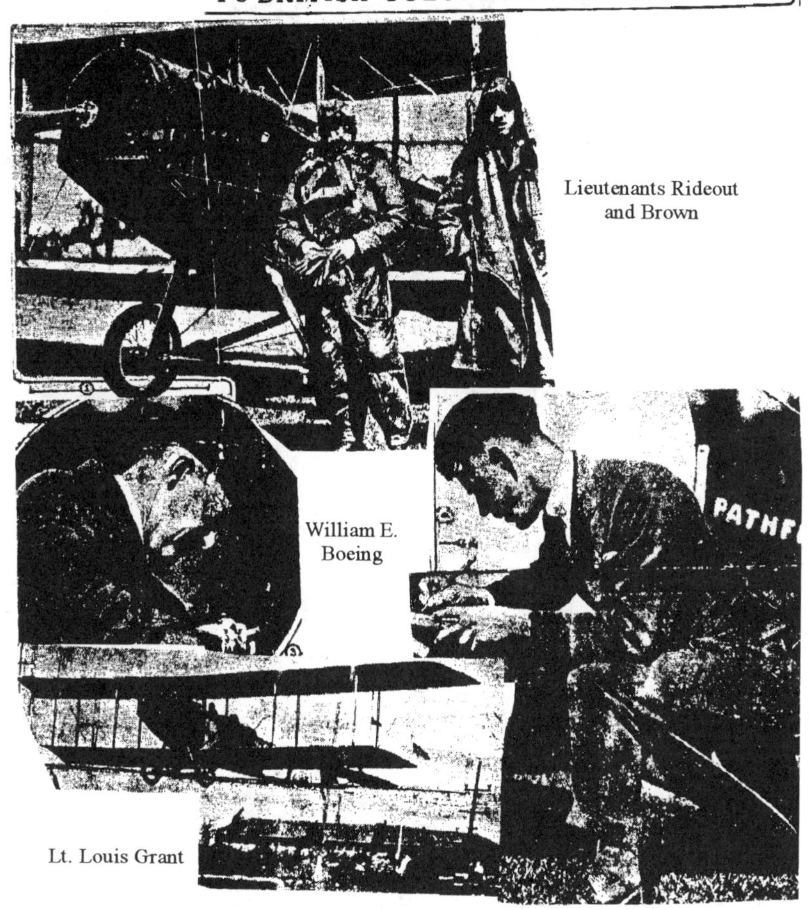

Lieutenants Rideout and Brown

William E. Boeing

Lt. Louis Grant

Eddie Hubbard went up to meet the Canadian airmen to lead them to the landing spot, as they dropped leaflets over the City inviting all the people of Seattle to attend the Victoria 24th of May celebrations. Lt. Louis Grant of the Aerial League had arrived in Seattle by ship several days before to make the Mayor and those interested in aviation aware of the flight. As a result Acting Mayor, W.D. Lane, Bill Boeing and over two thousand curious onlookers were on hand to greet the Canadian airmen.

Lt. Robert Rideout, a member of the Aerial League of Canada handing a letter from Victoria Mayor Robert J. Porter to Seattle Acting Mayor W. D. Lane beside Jenny "Pathfinder," May 18, 1919 at Jefferson Municipal Golf course. On the right is Rideout's flying partner, Lt. Harry Brown.

This initial invitation, in today's world, has resulted in not only participation by the Seattle area but also includes marching bands and floats from many cities in the states of Washington, California, Oregon, Idaho and Nevada. The Victoria Festival Society, who is in charge of the parade, estimate the annual economic contribution to the Greater Victoria area is about $500,000. Not a bad result started by two pioneer airmen dropping invitation leaflets over Seattle seventy-six years ago.

There were only a few airports in North America and any flat surface long enough to allow aircraft to take-off and land were put into use. This flight took off from the Willows Fair Grounds in Oak Bay and landed with the guidance of Eddie Hubbard at Jefferson Park Municipal Golf Course, Beacon Hill, Seattle and for those who pound the fairways this is the course where Fred Couples learned the game without having to jump out of the way of landing aircraft.

William E. Boeing handing letter to Lt. Robert Rideout for delivery in Victoria, B. C. after return trip from Seattle, May 19, 1919. Lt. Harry Brown is in the Jenny 'Canuck', christened "Pathfinder" two days earlier. This was the first North American International air mail to be flown from the United States to Canada. Two months previously Hubbard and Boeing flew the first North American International air mail from Canada to the United States.

The first North American air mail to travel from the United States to Canada went on the return flight. Acting Mayor W. D. Lane sent a letter to Mayor Robert Porter of Victoria. Bill Boeing and others took the opportunity to send letters to business and personal friends in Victoria. Seattle now became both the recipient and starting point of the first North American International air mail.

Rideout and Brown moved to California. Rideout became a stuntman and also played tough guys and gangsters at Columbia Pictures. Brown operated a flying school at Oakland until his retirement.

Chapter 6
Honeymoon Pilot

THE SEATTLE DAILY TIMES

Newly-Weds Start on Voyage to Cloudland

* * * * * * * * * * * *

Wings of Love Made Real by Seattle Pair

PRINCIPALS IN BEAUTIFUL ROMANCE OF THE SKIES

TO-DAY'S BRIDAL PAIR HONEYMOON IN A SEAPLANE

Miss Jessie Kennedy, of Colonist, Wed to H. S. Graves, of The Times

A flying seaplane was the novel and up-to-date form of honeymoon conveyance chosen by two popular members of the journalistic profession, Miss Jessie Elizabeth Kennedy, daughter of Mr. and Mrs. T. A. Kennedy, 94 Gorge Road, and Herbert Sandham Graves, son of the late Dr. W. H. Graves, F. R. C. S., of Dublin, and Mrs. Graves, of Rothwell Street, Victoria; whose marriage took place this afternoon. The bride has been the social editor of The Daily Colonist for some years, while the bridegroom is a member of The Times editorial staff. He saw much service during the war as a flight-commander in the Royal Air Force, and was one of the "originals" of the 16th Canadian Scottish. After transferring to the air branch of the service, he was attached for duty to General Allenby's Army in Palestine.

The ceremony was performed by the Rev. Dr. Sipprell, pastor of the Metropolitan Methodist Church, at the home of the bride's parents at 2.30 o'clock. The reception rooms were charmingly arranged for the nuptials, Shasta daisies, sweet peas and greenery being used with artistic effect, while in compliment to the bridegroom, the ceremony took place under a model aeroplane suspended

THE DAILY COLONIST, VICTORIA, B.C., SUNDAY, AUGUST 15, 1920

SEAPLANE HONEYMOON FOR VICTORIA COUPLE

Mr. and Mrs. H. S. Graves, Well-Known in Local Journalistic Circles, Married Yesterday, Take Novel Trip

The distinction of being the first bridal couple to leave Victoria on their honeymoon trip by seaplane fell to Mr. and Mrs. H. S. Graves, who were married in the city yesterday afternoon.

The marriage took place at half-past two at the home of the bride's parents, Mr. and Mrs. T. A. Kennedy, 95 Gorge Road, the principals being their youngest daughter, Jessie Elizabeth, and Herbert Sandham Graves, son of the late Dr. W. H. Graves, F. R. C. S., of Dublin, and Mrs. Graves, of Rothwell Street, Victoria. Rev. Dr. Sipprell, ... the Metropolitan Metho... performed the ceremony... witnessed by the ... of both families ... of the bride ... bridegroom ... mander ... during ...

Real Clouds for Seattle Honeymoon Couple

* * * * * * * * *

Heavenly Days Will Follow Pretty Wedding

... the bride's sister, ... of honor, wearing a ... of embroidered shell... with picture hat to ... she carried a bouquet of ... and pink sweet peas. The ... bridesmaid, Miss Elsie Smith, looked ... in a frock of white net with lavender-edged ruffles, the sash and underdress were of pale pink taffeta, and the picture hat worn was of white georgette garnished with French flowers. She, too, carried mauve and pink sweet peas.

The best man, Mr. Harry Brown, like the bridegroom, wore a full-dress uniform of the Royal Air Force. Mr. Brown served as a flight lieuten...

THE SEATTLE DAILY TIMES. FRIDAY EVENING, JUNE 13, 1919.

Pilot Edward Hubbard Puts in One Busy Day

* * * * * * * * * * * * * * * * * *

Finishes by Starting Honeymooners on Trip

SCENES BEFORE AND AFTER START OF BRIDAL PARTY FOR TACOMA

A series of honeymoon flights began with James Gibbs, a Fairbanks merchant, and his bride, the former Miss Vera Smith, using 'cupid' Hubbard to fly them from Seattle to Tacoma for their honeymoon. You may recall, Mr and Mrs. Gibbs were in the first group of passengers to use Eddie's Aerial Taxi. The happy couples flight took 25 minutes, at the rate of one dollar per minute.

Pilot Eddie Hubbard with honeymoon couple Mr. and Mrs. James Gibbs on their way to Tacoma, May 22, 1919. Twenty days previously the couple were among the first Aerial Taxi customers.

The second honeymoon couple, Ensign Elliott Dean Harrington, U.S.N. Air Service and his bride Catherine Butler approached Bill Boeing about a wedding trip in the clouds. Boeing agreed and put one of his planes at their disposal. The happy bride and groom left the Lake Union hanger for Tacoma at 6 o'clock in the evening after Eddie had flown his boss to Victoria, British Columbia, earlier in the day.

The third time Eddie played cupid was a year later, August 1920, when he flew to Victoria to pick up Sandy Graves and his bride. Graves had been a Royal Flying Corp. pilot in WWI and heard of Hubbard's exploits. Graves arranged for Hubbard to pick them up and fly them to Seattle. The Graves' hit the headlines as of the first bridal couple to leave Victoria for a honeymoon by airplane. This was such a rare event the departure from Victoria harbor was witnessed, not only by the wedding party, but also hundreds of other onlookers. At the time of his marriage Graves wrote a column for the *Victoria Daily Times* and later became publisher.

Chapter 7
First Foreign Aircraft to Visit Victoria

PREDICTS REGULAR AIR TRIPS ACROSS ATLANTIC SHORTLY

W. E. Boeing Addresses Luncheon at Empress Hotel on Aerial Development

AERIAL LEAGUE STARTS ITS DRIVE FOR FUNDS

SEATTLE SEAPLANE FLYING HERE TODAY

Mr. Boeing, Head of Plane Factory, Will Be Passenger and Will Attend Luncheon at Empress.

SEAPLANE ON VISIT HERE FROM SEATTLE

Made Trip Across in 69 Minutes; Is Popular on Sound With Honeymoon Couples

June 12, 1919 Eddie flew Bill Boeing to Victoria to address The Aerial League of Canada at a fund-raising luncheon held by the Rotary Club in the Empress Hotel.

Hubbard and Boeing left the hangar at Lake Union in the CL-4S seaplane at 9 o'clock, arriving at Cadboro Bay 69 minutes later. This was slightly faster than the Curtiss aircraft Rideout flew from Seattle the month before. Why Hubbard landed at Cadboro Bay and not Victoria's Inner Harbor, in front of the Empress, is a puzzle. The Canadian Customs officials were on hand to meet the visitors and inspected the seaplane by rowing out from shore. This was the first foreign aircraft cleared at the Port of Victoria.

Canadian Customs clearing Bill Boeing's aircraft at Cadboro Bay, near Victoria, June 12, 1919. This was the first foreign aircraft to land in the area. Hubbard was the pilot on this trip, bringing Bill Boeing to address the Victoria Rotary Club at the Empress Hotel.

Boeing's address was well received at the Empress and he made one of his usual interesting predictions. Within a very few years, he told his audience there would be large fleets of airplanes crossing the Atlantic. To prove his point he mentioned that ten years ago the first airplane flown by Bleriot crossed the English Channel. Alcock and Brown had just crossed the Atlantic a few days earlier. He was confident that in the next few years we would see the inauguration of regular service on the route. He urged the businessmen of the city to lend their support to the Aerial League of Canada, an organization attempting to establish and maintain aerial communication between different parts of the country. Eddie flew the boss back to Seattle, leaving Cadboro Bay at 2:30 in the afternoon. The luncheon speech at the Empress, once again, pointed out the accurate visions of William E. Boeing. With this type of leadership it is little wonder that the company bearing his name became a giant in the world of aviation.

Chapter 8
Eddie Rickenbacker Compliments Eddie Hubbard

Seattle Showers Honors on Captain Rickenbacker

* *

America's Ace of Aces Guest at Open Air Banquet

BANQUET TABLE, FROM SKY, AND CAPTAIN RICKENBACKER AT VOLUNTEER PARK

America's WWI aviator war hero, Capt. Eddie Rickenbacker, who shot down twenty-six enemy aircraft was given an outstanding welcome in Seattle during July 1919. An open air banquet at Volunteer Park attended by over 400 and a dance at the Masonic Temple were organized to honor the great man. Rickenbacker, who, at the time, manufactured automobiles and loved to race cars, had come to Tacoma to open the Speedway auto races and was *loaned* to Seattle for one evening. His party was brought to Seattle from Tacoma in Bill Boeing's yacht, the Taconite, and among his racing compatriots were people such as Cliff Durant and Louis Chevrolet familiar names in auto manufacturing.

The Flyers' Club had organized the banquet and arranged for Eddie Hubbard to pick up the guest of honor in Tacoma. Unfortunately a broken compression strut, which took two hours to repair, prevented Eddie from having the pleasure of flying one of his idols. Hardly had the banquet started when Hubbard appeared with E. Pierson & Co., photographers to take pictures of the affair. Hubbard then dropped Pierson off at the Lake Union hanger and returned to put on a dazzling display of aerobatics that gave the huge crowd a tremendous thrill.

Rickenbacker led the crowd to its feet and started a roar of applause. Turning to a table companion he said,

> "Hubbard enjoys a great reputation among airmen everywhere. Many who have taken instruction from him believe he is the most skilled pilot in America. I can readily see where he gets that reputation. He's a corker. They don't do that stuff any better anywhere."

What better compliment could ever be bestowed upon Edward 'Eddie' Hubbard, the Northwest's outstanding aviator, particularly when it came from such an accomplished aviator as Rickenbacker.

Chapter 9
The Popular Aerial Taxi

Leaves by Plane to Catch Kashima Maru for Siberia

DAILY AERIAL TAXICAB SERVICE STARTS IN SEATTLE

It seemed that Hubbard was on the go all the time. A few days after Rickenbacker's visit, a Red Cross official had missed the steamship Kashima Maru heading out of Seattle for the Orient. Someone suggested the only man in town that could catch the ship was Eddie Hubbard. He was contacted at the hangar and flight arrangements were made. With his passenger strapped in, Eddie took off and overtook the ship at Port Townsend. He landed on the ocean and put the Red Cross official on a tug that intercepted the steamship. This was the first time an airplane had overtaken a ship to put a passenger onboard.

Eddie's name was becoming a household word in Seattle and Victoria. There was hardly a week went by that his name or a picture of him at the controls of the Boeing plane was not in newspapers.

BOEING PLANE CARRIES P.-I. WRITER TO VIEW TRUCK PARADE

More business for the Aerial Taxi came from *The Seattle Post-Intelligencer* when they sent one of their reporters up with Eddie to report on a huge truck parade. The reporter was so impressed by the flight he ran out of adjectives.

May 1929 -- Eddie decided to buy a car and came up with a unique way to determine its power and speed. He told the salesman to go to a point on the bridge near the hangar at Lake Union. Eddie went up in the Aerial Taxi and raced the car for a short distance. Our pilot must have been satisfied with the performance because he took delivery of a new car from Cascade Motors.

Make Flight Over City in Boeing Taxicraft

* * * * * * * * * * * *

Now They Want to Own Their Own Planes

EMBARKING ON THEIR FIRST AERIAL TRIP OVER SEATTLE

Boeing almost made a sale when two ladies took the Aerial Taxi for a spin around the town and were so enthusiastic they wanted to buy planes for their own use!

Clubs in the area started using the Aerial Taxi. One on Bainbridge Island sent two of its members to have a ride from Lake Union back home. Not to be outdone, the Bremerton Yacht Club sent their best dancing couple at its weekly "hop" to the Boeing hangar for a ride.

Commercial firms started taking advantage of the taxi when a Seattle advertising firm, Strang & Prosser, chartered Hubbard and the Boeing B-1 to take aerial photographs of new subdivisions. Realtors claimed it was far easier to show a realistic view to potential buyers without even visiting the site. These early promoters of aerial photography went on to say this is just the beginning. How right they were!

Hubbard preparing for aerial photography with Frank Jacobs.

The headlines read,
"TIRES DROP FROM SKIES IN EVERETT."

When Eddie 'Bomber' Hubbard demonstrated the advantages of delivery by air without even landing. He flew a shipment of tires on the CL-4S seaplane from Seattle to Everett and arrived less than forty-five minutes after the order was placed. He simply strapped seven tires under the wing of the plane. His aim was so accurate the tires landed in a vacant lot across from the tire store. He released the tires by cutting the rope that held them in place under the wing. After receiving a check for the shipment, Boeing Sales Manager Berg sent The John K. Healy Co. a letter commending them for their aggressive and progressive business policy in employing this most up-to-date means of transportation for a rush delivery. This was another of the many firsts in Eddie Hubbard's career. A few months later the Diamond Rubber Co., Inc., of Ohio placed a full page ad in *Hardware Age* magazine describing this unusual event.

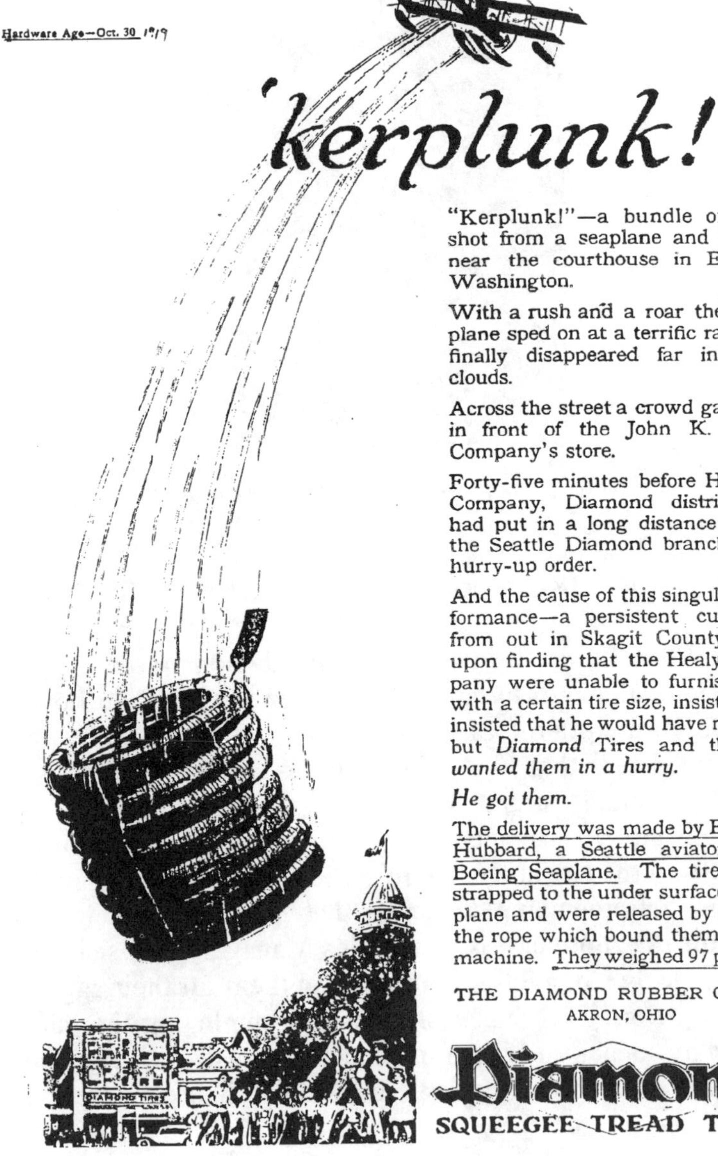

BOEING AIRPLANE COMPANY
SEATTLE, WASHINGTON
GEORGETOWN STATION

October 8th, 1919.

The John K. Healy Co.,
Everett, Wash.

Gentlemen:-

Your check to our aerial Transportation Department covering the delivery charge for a shipment of Diamond tires, via Airplane, from Seattle to your store in Everett, has been received, for which we desire to express to you our sincere thanks.

We wish to take this opportunity to commend your aggressive and progressive business policy in employing this most up-to-date means of transportation in order to make a rush delivery.

This is the first time to our knowledge that such a shipment has been expressed or delivered by Airplane, and besides having made history in this connection in the North West, you have undoubtably established a precedent that gives a new definition to the word "service."

Assuring you of our keen appreciation of your desire to co-operate in promoting Commercial Aviation, we are

Yours very truly,

BOEING AIRPLANE COMPANY

HCB/F

H C Berg

Sales Manager

BOEING AIRPLANE COMPANY
SEATTLE, WASHINGTON
GEORGETOWN STATION

September Tenth,
Nineteen Nineteen

Mr. W. E. Boeing,
1100 Hoge Bldg.,
Seattle,
Washington

Dear Sir:

There is quoted below letter received from the Bank of Stanwood, regarding our exhibition flight over Stanwood on the 5th inst.

"I am enclosing to you herewith our Treasurer's check for $350.00 in payment of your account with our Committee in sending the machine to Stanwood for the Home Coming celebration.

"Permit me to express to your good self and Company our appreciation for your service in this connection and the splendid exhibition given by your pilot.

"His arrival was timed perfectly and we hope that he reached home all safe. Will you kindly give him my personal thanks for his part in the days grind for he no doubt can appreciate the effort to please a multitude on a day like this. With personal regards I am

Yours very truly,"

(Signed) O. E. Thompson
Cashier

EMG:BB
COPY TO MR. HUBBARD

Notify Hubbard

RECEIVED
SEP 13 1919

Eddie was bringing in a good deal of money as noted in a letter to Boeing from Gott dated September 10, 1919. The company received $350 for Hubbard putting on a show at Stanwood's Homecoming celebration. Bill Boeing initialed the letter and penciled, "Notify Hubbard."

SEE THE FLEET
FROM A
BOEING
SKY TAXI
and dip and float like a lazy cloud—over the Pacific Armada in a Boeing Sky Taxi. Inspect Uncle Sam's floating fortresses—or see the presidential review, from the sky. An honest - to - goodness safe and secure thrill for sensation-jaded systems or red-blooded men and women who appreciate the uncommon. Big, comfortable Boeing Aircraft will make regular flights over the fleet from our Lake Union hangars. Skilled, highly experienced pilots in charge. Enjoy fifteen minutes of real novelty in the air for only fifteen dollars—a trip your friends will envy. Only a few reservations left. Get one of them by phoning Capitol 758.

In September the Fleet came to town and headlines screamed: "ARMADA IS HERE."

Contracted by F.W. Strang of Strang & Prosser, Hubbard in the B-1 flew Strang and photographer E. Pierson over the flagship New Mexico, one of four battleships, in the fleet of twenty-seven vessels which were just arriving from Victoria, B.C. Not only did the flight provide many aerial shots of the fleet entering Seattle's harbor it also gave the men of the New Mexico an early delivery of the morning newspaper. Hubbard dropped a bundle of *Seattle P.I.*'s on the deck of the battleship.

AIR MACHINE WAS REQUISITIONED TO DELIVER GLASSES

Nothing was impossible for Eddie Hubbard. In the excitement of departure aboard the Suwa Maru, from Seattle, a lady dropped her glasses overboard. The loss was noticed by her friends on the dock and they took immediate action by sending a telegram with her prescription to Victoria, the ship's next and last stop before heading to the Far East. Her new glasses were waiting in Victoria as the liner docked however, taking no chances, the prescription had been filled in Seattle and her friends contacted Eddie to fly after the ship. He overtook the liner off Port Townsend and dropped the well wrapped package onto the ship's deck. Another 'bombing' right on target.

For the annual football game between Washington and Oregon Eddie flew Miss Mildred Jackson, a University of Washington senior, over the stadium. Miss Jackson dropped the ball to open the game. Reports do not indicate if it was caught.

EDDIE HUBBARD RUNS FIRST AERIAL FREIGHT SEATTLE TO TACOMA

City of Destiny Gets Its First Electric Clothes Washing Machine

With a washing machine sitting on the lower wing of the CL-4S, Eddie takes off for Tacoma to make a delivery in the Aerial Taxi. The load was so heavy he couldn't lift off from small Lake Union. He eventually taxied to Lake Washington and made the lift-off.

The last of Hubbard's accomplishments for 1919 was another freight delivery. He flew a 400-pound washing machine to Tacoma and almost disappointed the many spectators when he tried from 12:30 until 2:50 p.m. to get the seaplane off Lake Union. Finally he taxied to Lake Washington and with stronger winds was successful in getting airborne. He ran out of gas just short of Tacoma and a boat came to his rescue with enough gas to complete the journey. Mrs. Gott, wife of Boeing's General Manager, suggested to Eddie he should do his washing on the way and hang it up for fast drying. After finding a fuel supply Eddie was on his way back to the Lake Union hangar.

THE SEATTLE DAILY TIMES, FRIDAY EVENING, JANUARY 2, 1920.

Too Cold for Tetrazzini to Fly
* * * * * * * * *
Dora Dean Takes Diva's Place

Reminded by Entourage of Danger to Her Throat, Singer Calls Off Trip and Times Writer Goes Up Instead.

The Boeing Aerial Taxi had become so well known that a visiting concert singer, Madame Tetrazzini, had her manager book a flight with Eddie. Madame arrived at the hangar with the manager, maid, violinist, manager's wife and the violinist's wife. There was a cold wind that day and her manager and the rest of her troupe begged Madame not to fly. She had a concert to give that evening and the cold would harm her delicate voice. Finally she gave in to the pressure. *The Seattle Daily Times* had sent one of their reporters, Dora Dean, to cover the event. The flight was already paid for so Eddie suggested Dora take the great singer's place. Dora's write-up gave a glowing report on the 50-minute flight. Her final line, "It was grand!"

Eddie helping child into Boeing B-1 cockpit at Redondo Beach company picnic.

Two more international firsts were chalked up by Eddie Hubbard. He flew his boss, Bill Boeing, to Vancouver to take part in a golf tournament. They no sooner returned when a frantic phone call was received. Judge Frederick Brown of Seattle was recuperating in Victoria from an operation he had in St. Paul a few months earlier. He suffered a relapse and a Victoria physician contacted his wife at 3 a.m., suggesting she get to Victoria as soon as possible. Mrs. Brown immediately called and arranged for the Boeing Air Taxi. Along with her daughter, Jessica Ross, they left Seattle at 7 a.m. and pilot Hubbard landed an hour and fifteen minutes later at the Royal Victoria Yacht Club, Cadboro Bay. Within 15 minutes the wife and daughter were at the judge's bedside at the Empress Hotel. Unfortunately he passed away that afternoon.

HASTEN TO DEATHBED IN BOEING SEAPLANE

Trip From Seattle to Victoria Made in Fast Time

VICTORIA, July 7.—The death took place in Victoria yesterday of Hon. Frederick V. Brown, of Seattle, general solicitor for the Great Northern Railway Company, Western Division. He was 53 years of age.

In the later part of August 1920, Bill Boeing took his friend J.T. Keena trout fishing. They left Seattle in his yacht, *Taconite*, and picked up J.A. Rithet in Victoria. They went north and anchored off Campbell River. In no time the *Taconite* was surrounded by local indians who told them there was a lake over the hills that was very hard to get to but the trout were so plentiful they came up and bit your hand! The lake was Buttle Lake and for Boeing there was a quick, easy way — call Eddie. The next day he arrived in the B-1, surprising all the local residents. The *Comox Argus* newspaper announced this was the first plane to visit the area. After a day and a half the Boeing party left the lake and flew back to the *Taconite*. Eddie took the trout to Seattle. The fishing was so good observers said the B-1 had a starboard list from the weight of Buttle Lake trout!

Bill Boeing's August 1920 Canadian fishing trip to Campbell River, Vancouver Island, B. C. Boeing and party arrived in his yacht. Eddie flew the B-1 from Seattle and transported the anglers over the hill to Buttle Lake for a once-in-a-lifetime trout fishing experience. The B-1 is at anchor in front of the original Willows Hotel.

B-1 at Campbell River, B. C., on Buttle Lake fishing trip August 1920. Boeing's yacht *Taconite* is on the left. Note smoke coming from coal-burning tug.

B-1 at Buttle Lake, B. C. fishing trip.

In the meantime Eddie was still kept busy doing flying jobs around town. Paramount Pictures wanted to publicize National Paramount Week at all their theatres. Eddie and the Aerial Taxi were hired and flew one of the theatre managers over the downtown area dropping advertising leaflets and passes.

Flying is a serious business; however, it does have its humorous moments. When Eddie rturned from the successful fishing trip the Boeing Airplane Company received a very odd letter. The letter on the following page is reproduced from Boeing Archives.

Seattal Wash

Sept 30 1920
Boeing Airplane Company
Dear Sir:-
Can you give me aney advis as to Hoom I mite Communacate
With In Regard to flying I Dont feal as thoe I can pay a Dollar a minuit.
fot I Wood Probly Haft to Have Considerbal training Bee for I Wood Bee abel to handle a Mashean a Lone if you can give me the Names of Some airvator Ore aney one in the bisnis will Preashate it.

Yours Truley

October 6th, 1920

Dear Sir;

Replying to your inquiry dated September 30th, suggest that you communicate with Mr. Edward Hubbard, Ft. of Roanoke Street, Seattle, Washington and ascertain his rates for instruction
You might also get in touch with the Leshi Aerial Taxi Company, Leshi Park, Seattle, Washington and find out if they give instructions and their rates.

Yours very truly

Edgar N. Gott
General Manager

ENG:BB

I wonder what ever happened to this budding pilot?

Eddie came to the rescue of *The Victoria Times* by flying in twenty-five dry molds necessary for the production of the next edition. It was reported that this was the first time in Canadian history that a newspaper had used the services of an aircraft in order to keep the presses rolling. Even the *Victoria Colonist* reported on their opposition mentioning that a strike at the Victoria Gas Company caused a serious problem in the stereotype plant of *The Times.* Dry molds had to be used instead of gas molds. The dry molds were lent to its Victoria namesake by *The Seattle Times.*

The French hero, Marshall Joseph Joffre, landed at Victoria and *The Seattle Times* beat all competition by hiring Eddie to fly photo plates of the great man to Seattle. *The Times* had their own photographer and a staff correspondent in Victoria. As soon as the plates were ready the correspondent gave them to a waiting Eddie. As soon as he docked at the Lake Union hangar a *Times* automobile rushed the plates to the newspaper office for development. *The Times* proudly displayed the picture of Marshall Joffre which was printed in the paper the same day it was taken across the international boundary. What a small world it had become!

The Boeing Airplane Company finally entered into a lease with Hubbard for the hangar on Lake Union at $50 a month.

In June the *Seattle P.I.* had a picture of Eddie in his B-1 flying boat pulling two fair maidens on a board at Lake Washington. Maybe this was the start of water skiing.

In the summer of 1922 the fleet sailed into Elliott Bay and again Eddie was called to fly the *P.I.* photographer aloft to have the event covered for posterity.

For the first time in his life Eddie was in the glue. The University of Washington homecoming committee approached him to fly two co-eds over the campus to distribute handbills to advertise the opening celebrations the next day. This idea didn't sit well with the editor of *The Seattle Times,* which had a very strongly worded editorial outlining the dangers of an aircraft flying low over a large crowd. The flight was called off with no comment from either party.

The Aerial Taxi was at it again. The President and Secretary of the Inglewood Country Club were flown from Lake Union to the golf course on Lake Washington by Eddie on a Sunday to start the day's golf. *The Times* had almost a whole page write-up along with photographs of the event.

The Aerial Taxi also came to the rescue of two young people and a dinner date. It seems that Donald and Loretta Haines were to travel to Victoria to meet the incoming liner, *President Madison*, and dine aboard while the ship travelled to Seattle after clearing Victoria. However, they missed the C.P.R. ship to Victoria and that was when their father suggested taking an airplane. Eddie was travelling to Victoria later in the day to pick up mail from the *Madison* so it was no trouble to take the two youngsters along. He did it in his usual efficient manner and deposited his passengers at the Victoria Outer Wharf dock before the President Madison tied up.

Eddie's second taxi competitor began business in Seattle. The first competitor was Herb Munter with a land plane operating out of a field in Kent. Munter wasn't that much opposition as he specialized in sightseeing trips to Mount Rainier an area Hubbard did not take the seaplane or flying boat. Now another air taxi service, Puget Sound Airway Company, started operating with a seaplane owned by Leo Huber. His company had three aircraft. One was a Curtiss Gull and the others were Boeing seaplanes operating out of a hangar at Madison Park on Lake Washington.

Visiting Noble "Arrested" And Seattle Shriners Rush to Aid

Harry Janes, a Shanghai shriner, on his way to a convention in Seattle, jumped ship at Victoria, contacted Eddie Hubbard and flew to Seattle, arriving ahead of his group from Shanghai. He talked police chief Bill Severyns into "locking him up" and waited for his pals to get him out of jail.

Eddie was to play a part in some Shriner hijinks. A group of Shriners arrived in Victoria from Shanghai on their way to a convention in Seattle. One of the group, Harry Janes did an about turn on his buddies, jumped ship at Victoria and had Eddie fly him to Seattle. As soon as they landed Harry visited Seattle's finest and cooked up a scheme with Bill Severyns, Chief of Police. When the *President Jefferson* arrived from Victoria with its Far East passengers the local Potentate was informed that one of their visitors from Shanghai was being held without bail at the city jail. This put all the Seattle Shriners into a panic with the local Potentate breaking all speed records getting to the city jail. They found their 'brother' in custody; however, the police officials refused to be swayed by their pleas to release the 'prisoner' in their custody. For an hour many prominent Seattle men argued for Harry's release and the city jail looked like a Shriner's convention. Finally, the "prisoner" was released on a solemn promise that he would be delivered back to them in the morning. At the evening banquet in the Olympic Hotel his 'brothers' realized they had been had!

An interesting article appeared in the *Seattle P.I.* with the headline 'First Aerial Taxi Service In U.S. Will Start Operation In San Francisco.' It was by a special correspondent of Universal News Service. The *P.I.* staff goofed on this one. Eddie and Bill Boeing had the first Aerial Taxi on the West Coast and possibly the first in the U.S. in 1919, six years previously.

Eddie's B-1 Aerial Taxi flying boat was in the news once more with a half page picture in *The Times* showing his passengers, on a flight to Port Angeles. They were Colonel and Mrs. Drain and their son. Colonel Drain, from Washington, D.C., was National Commander of the American Legion.

SEAPLANE CARRIES HUSBAND IN PURSUIT OF FLEEING WIFE

Victoria Times Oct 7, 1925

Irate Japanese Overtakes President Jefferson by Air and Demands Wife; Brother of Woman Borne to Sea Against His Wish.

Pilot Hubbard and the Seattle-Victoria aerial mail plane played-

Many times Eddie and his mail plane had played cupid to honeymoon couples. Here is a story with a different twist. A Japanese resident of Seattle hired the Aerial Taxi to fly to Victoria so he could 'rescue' his wife and one child. It seems his wife had a touch of homesickness and left Seattle on the *President Jefferson* for the Orient, via Victoria, with her visiting brother. Her husband, left at home with three children, was to say the least, a little upset. Eddie had him at the Victoria dock long before the Jefferson arrived. The net result was that the wife and child were held at the Canadian Immigration Building overnight while her brother, not allowed to embark, sailed for his home in Japan. The next day, husband, wife and child returned to Seattle. I wonder if they lived happily ever after!

Eddie and his mail plane took part in a baseball joke. The *P.I.* ran a picture of the flying boat with one of the coast league outfielders in it. The reason - Ike Boone was hitting so many home runs the Seattle outfielder said he needed an airplane to catch the balls coming his way. I wonder how many he caught!

Hubbard approached Boeing President Gott and General Manager Phil Johnson regarding construction of a new flying boat utilizing the Liberty engines he had on hand. Eddie also wanted the outrigger tail type, a monoplane with a span not to exceed 55 feet, loading not to exceed 8 pounds to the square foot and to carry a payload of 800 pounds. His B-1 had a span of 50 feet. Designing the new flying boat was put on hold because the engineering department was in the process of designing a new pursuit ship. The plant remained busy, so Hubbard forgot the idea of a newly designed flying boat and had the B-1 modified with new wing pontoons and a radiator shell and shutter installed.

Chapter 10
San Francisco Aeronautical Show

BOEING AIRPLANE CO. TO SEND EXHIBITS TO SHOW

Local Aircraft Builder Will Make Display That Will Equal Best—Have Special Features

Boost for Seattle in Aircraft Exhibition

The first postwar chance the Boeing Airplane Company had to show off its capabilities happened was at an Aeronautical Show, April 1920, held under the sanction of the Manufacturers' Aircraft Association in the San Francisco Civic Auditorium.

Both Boeing and Hubbard attended the show and took two flying boats, the BB-1, B-1, plus a land plane and a sea sled. The BB-1 had a passenger capacity of three with a 125 h.p. Hall-Scott engine and a speed of 85 miles an hour. The B-1 had the capability to carry five passengers with a 200 h.p. Hall-Scott and a speed of 90 miles per hour. The land plane was designed for Herb Munter, who was now flying his Aerial Taxi from the field in Kent.

Eddie Hubbard about to land in the BB-1.

Eddie about to take two damsels aloft in the BB-1. Note the ladies' leather coats, helmets and goggles while Eddie is in street attire.

The show lasted a week and was a great success. Boeing's exhibits received some very favorable comments. Although the show was not intended to sell machines Boeing made the only sale: a 24-foot sea sled to a lady from Lake Tahoe.

Following the show both Boeing and Hubbard went to San Diego for three months to carry out tests at the Naval Station. Hubbard was able to demonstrate that the BB-1 was well in advance of anything yet developed in the line of flying boats. He flew just about every naval pilot on the station in the new plane and received enthusiastic response on its behavior. Eddie also made a survey of passenger handling in the South and found that the comfort and safety of passengers were much better in Seattle. He also observed that California sightseers were taken up only about 1,000 to 1,500 feet, while he went up to 3,000 feet, giving his customers a range of vision of fifty miles or more.

If Hubbard coughed, the papers reported it. Back in Seattle after his three months in the South he had to get the Aerial Taxi business off the ground again. To obtain publicity he gave an interview to the press which is summarized as follows:

"Eddie stressed the importance of keeping aircraft in perfect condition and eliminate deterioration by using a hangar. He also plugged the 3,000 foot altitude for sightseers and the aircraft he used were built for customer comfort. He also touched on the future of air mail which was to later become a big part of his life. Eddie predicted aircraft with enclosed cabins and again greater passenger comfort. His predictions were just around the corner."

Chapter 11
Seattle's First Air Field

Hubbard had been actively promoting two important aspects of aviation, landing fields and air mail. During his trip to California he mentioned the number of good landing fields in that State. Bill Boeing presided at a conference sponsored by the Seattle Transportation Club. The sole purpose of the meeting was to start the ball rolling for a landing field in the city limits. At this time there were only two landing spots available: The Jefferson Municipal Golf Course and Munter's field in Kent. The golf course was only a temporary situation and was really too small. A crash, on takeoff, had already occurred. Munter's field was too far out of town. By 1920 the County Commissioner was concerned about Seattle being left out as an air mail center. Already Congress had given appropriations to Portland and San Francisco for airfields. Finally Seattle's hopes were raised when Congress adopted a resolution pledging the Government's acceptance of the Sand Point site on Lake Washington.

AVIATION FIELD DEDICATION HELD

Opening of Flying Base at Sand Point Marks New Era for Seattle.

CEREMONY IMPRESSIVE.

Cross Country and Transpacific Air Route Terminal Is Prediction.

In May 1920 Otto Praeger, United States Assistant Postmaster General arrived in Seattle to address a Chamber of Commerce luncheon and urge the establishment of a landing field. To make his point he told of the banks in Minneapolis and St. Paul that were saving more than $1,000 a day in interest as a result of air mail service. He added that cities must cooperate with the Post Office in making landing fields available. Those with fields would be given first consideration and Sand Point would be an ideal location for Army, Navy and air mail use.

The need for a landing field in Seattle again hit the headlines. An Army pilot flying from Eugene, Oregon was told to land at Kent Field, south of Seattle. He missed the field and running short of fuel had no other choice than to use the Jefferson Municipal golf course. The pilot knew two things about the course. The golfers didn't appreciate the fairways being torn up and it was really too small an area to land safely. The red faced pilot apologized for ripping up the ninth green and bending the flag pole with his wing! He finally came to rest, without injury, between the starter's house and a bench.

Things were starting to move. The County bought 219 acres at Sand Point for a landing field and the Post Office officially announced air mail service between Seattle and Victoria had been approved. Negotiations with Canadian authorities would begin immediately for approval to pick up and deliver mail at Victoria, British Columbia.

On June 19, 1920 a dedication ceremony was held at Sand Point. Eddie flew the County Commissioner, Claude Ramsay, from the Lake Union hangar to the ceremony in Bill Boeing's seaplane. After Boeing and the three County Commissioners turned over the first soil Boeing remarked that the plane in which Hubbard had flown Commissioner Ramsay to the future flying field, was the same aircraft that carried the first North American International air mail from Vancouver, B.C. to Seattle the previous year.

The most unpopular man in Seattle was County Commissioner Frank Paul. The Army asked for $500 to drain Sand Point landing field which the Government was leasing from the county. The Army wanted the field in good shape for the "around the world flyers" starting from Seattle in two weeks. For appropriation of county funds all three commissioners had to vote in favor. Two were in favor. Paul said, "I fail to see where the Board could truthfully declare that an emergency exists and make an excuse for an illegal expenditure."

The Seattle Times came to the rescue and announced in a very large print editorial, "SAND POINT WILL BE READY." Their community minded $500 donation saved Seattle from an embarrassing situation.

Edgar Gott, Boeing Airplane Company Manager, addressed the Seattle Chamber of Commerce and had this to say to local businessmen.

> "Seattle now has the largest aircraft manufacturing plant in America. Eddie Hubbard's air mail route to Victoria has become the great example of how planes can handle the mail. Now 50 per cent of the commercial air mail load in the entire country travels by air between the two cities." Gott made a some good points in his address. Comparing types of transportation, he noted the automobile was subsidized by the construction of roads and it was about time the Government did what European governments were doing for aviation - build airports. He also advocated a system of examinations and licenses for airline pilots. He made his point by saying, "Aviation is the only form of transportation besides shoe leather that does not require a license!" Actually he forgot horses.

Gott left Boeing and returned to Seattle as Vice President of the U.S. Fokker Corporation headquartered in New Jersey. He was promoting the future of commercial aviation but made his point by saying the country desperately needed more landing fields. He also made Eddie Hubbard the West Coast dealer for the Fokker Corporation. *Aero Digest* mentioned two Fokker Universals had been delivered to Mr. Edward Hubbard of Northwest Airlines Inc., Seattle, Washington and created quite a sensation on the Pacific Coast. The arrangement didn't last long and from then on Eddie stayed with the aircraft he was used to - Boeing.

Edgar Gott, Boeing Company manager left to become the U. S. Fokker distributor. He made Eddie the West Coast dealer out of Seattle. This arrangement lasted only a short time.

The actual cost to remove the timber and level eighteen acres at Sand Point in 1920 was $165,000. With the price of logs in the present market they would have made an estimated $200,000 profit!

Continual promoting by Eddie Hubbard and Bill Boeing for a city landing field in Seattle finally paid off. Sand Point was turning into a strictly military field and the use of it by other than military aircraft would soon come to a close. The Board of County Commissioners authorized the purchase of 425 acres at Georgetown in south Seattle for a commercial airport. Once again Commissioner Paul threw a monkey wrench into the proceedings. After the Board had been working on acquiring the site for some months, Paul at a meeting that was to have finalized the site acquisition left abruptly. Before leaving he said he favored the airport so long as it didn't cost anything. Eventually with state financing the airport construction went ahead.

Chapter 12
Canadian Crash Inquiry

FOR TECHNICAL SCHOOL

The Vancouver Sun

THE WEATHER

THIRTY-SIXTH YEAR—No. 231 VANCOUVER, B. C., THURSDAY, AUGUST 19, 1920 5¢ A COPY FOURTEEN PAGES

VANCOUVER FLYER FALLS TO DEATH

Likely to Hold B.C. Referendum On October 25

This Will Be Date If Voters' Lists Are Ready

Other Provinces to Take Their Votes Same Day

Government Anxious to Have Fair Decision

ANGLO-JAPANESE ALLIANCE OPPOSED

UNIONS ENTER STIFF PROTEST

SOLDIERS' SETTLEMENT

WEST ORIGINATES MOST OF TRAFFIC ON RAILWAY LINES

EASTERN EXPENSES SHOWN TO BE HIGHER

NATIVE SON KILLED IN FLYING CRASH

Warsaw Now Likely Saved From Russians

CAPTAIN BRENTON MEETS TRAGIC END IN ENGLISH BAY

Purchasing Agent for Union Steamship Co. of B. C. and Former War Aviator, Victim of Vancouver's Second Air Tragedy of Year

THOUSANDS SEE MACHINE DROP; BODY NOT FOUND LAST NIGHT

Pilot Was Using Boeing Flying Boat in Attempt to Gain Commercial License—Relatives of Deceased Witnesses of Dive to Death

In July 1920 Eddie delivered a new Boeing BB-1 flying boat to the Aircraft Manufacturing Company in Vancouver, B.C. One of this company's owners, Capt. Hoy, was the first to fly over the Rocky Mountains from Vancouver to Calgary a year earlier.

The Boeing BB-1 flying boat crashed with Capt. Hoy at the controls. It was not serious and no one was injured. A month later the BB-1 crashed again with very serious consequences. A young Canadian Navy pilot, Captain Hibbert Brenton fell out of the aircraft to his death at English Bay in Vancouver, B.C. A Court of Inquiry was set up in Vancouver at the end of August and at their request Eddie flew from Seattle to be a witness and gave the following testimony:

Question: How long have you been flying Boeing seaplane machines?
Answer: I had about 1,400 hours over a period of about two years.
Question: How long with this particular type of machine?
Answer: About 30 hours covering a period of about three months.
Question: This particular machine in question, was she a very sensitive machine on controls?
Answer: She was for a water-type machine.
Question: Had this machine any characteristics?
Answer: One of the most characteristic things of this flying boat is that the location of the centre of thrust is slightly above the centre of gravity and in case the motor should suddenly be cut off, the machine would automatically become tail heavy and have a tendency to rise suddenly in the stall. The machine would then whip with enough force to throw a man out who was not strapped in, otherwise I do not know of any position that the machine would naturally go into that would tend to throw the pilot out, providing he was not strapped in.
Question: Would she fly should the ailerons be out of commission?
Answer: Absolute control can be kept of the machine and it can be maneuvered very well without the use of the ailerons.

After another eleven witnesses The Board of Inquiry came to the following conclusion:

No evidence can be obtained as to what actually happened but it is possible for one of the gloves which witnesses say were thrown by the pilot into the cockpit before starting the flight and afterwards found that one of them was very badly ripped - in the water, to have jammed the pulley between the passenger seats which carries the elevator control wires, thus temporarily jamming them.

The Board also made recommendations to prevent a repetition of this type of accident:

To make it compulsory that no pilot who has not flown for over six months be allowed to fly solo until instruction has been given dual control for at least two hours.

That safety belts must be fitted to all machines - private or commercial - for the use of pilots and passengers.

Chapter 13
United States First Foreign Air Mail Contract

A day after Assistant Postmaster General Otto Praeger's May 1920 speech in Seattle urging the city to come up with a landing field, he, Seattle Postmaster Edgar Battle and Edward McGrath, Superintendent of Railway Mail Service discussed the possibility of a hydroplane service between Seattle and Victoria, B.C., which would expedite mail to the Far East and bring incoming mail from the Orient to Seattle. McGrath noted the saving of a day on delivery of commercial paper would mean an important financial advantage to business firms. Boeing Vice President Edgar Gott confirmed the Post Office officials' statements saying that hydroplanes were available with a mail carrying capacity of 750 pounds.

Praeger was a great promoter of air mail's future and to convince his audience he added that air mail service, in its second year in the East, demonstrated it was dependable and superior to the train. The cost was not excessive when its value to the community was considered and the New York to Boston route was saving the post office $120,000 a year; the New York to Cleveland route, $160,000 a year.

Seattle and Victoria newspapers featured front page reports on the possible air mail service between the two cities. This news was exactly what someone else was waiting to hear - Eddie Hubbard, a test pilot for the Boeing Airplane Company.

In the meantime Hubbard was test flying the new Boeing land plane built for Herb Munter and his company, Aerial Tours. Boeing officials and Hubbard were more than satisfied with the test results. Without any difficulty Eddie took the machine up to 17,000 feet with two passengers. It hit a speed of 100 miles an hour. This was the plane that had been a big hit at the recent San Francisco show.

Encouraging news was received from Canadian Postal authorities regarding the handling of U.S. mail arriving at Victoria on Canadian Empress ships. Edward McGrath, Superintendent of U.S. Railway Mail gave this statement, "Our plans are moving slowly, but I am sure it will not be long before Seattle will have the distinction of being the first city in the United States to receive international mail by air."

NIGHT EXTRA

The Seattle Daily Times

SEATTLE, WASHINGTON, SATURDAY EVENING, OCTOBER 2, 1920

SEATTLE AWARDED FIRST INTERNATIONAL AIR MAIL SERVICE

AERIAL DELIVERY TO SPEED LETTERS ON WAY TO ORIENT

AERO MAIL FROM SOUND ASSURED

Definite Announcement of Service; First of Kind on Continent

MAIL PLANE LEAVING WATER

Seattle-Victoria Air Mail Service Starts.---Meets Oriental Steamers

This is what Eddie Hubbard had been waiting for. It was now time for him to submit a bid. He was not the only one interested in the proposed air mail route. Another Seattle company, Leschi Aerial Taxi, did a trial run from Seattle to Victoria and submitted a bid.

October 2nd Postmaster General Burleson announced the mail contract had been awarded to "Edward Hubbard, Contractor," and service would begin October 15th. The news was carried by almost every newspaper in the state of Washington and many elsewhere. *The Victoria Times* had an editorial announcing the good news and it ended with a very prophetic statement, "In any case it is plain enough that Hubbard's contract is merely the forerunner of what will become a very general business all over the world before long, despite all pressure or opposition exercised at official quarters."

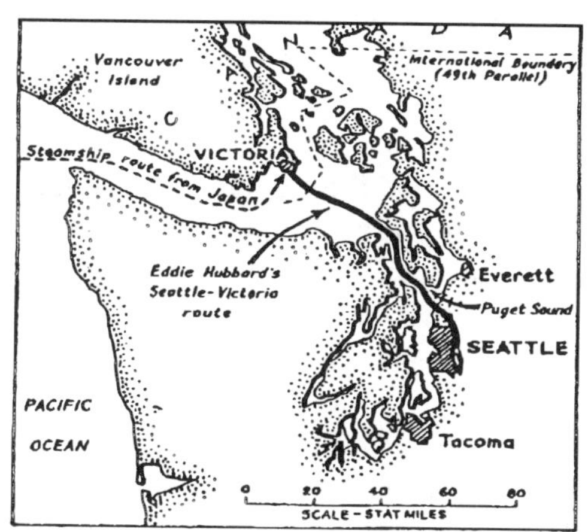

Map of the route operated between Victoria and Seattle

The United States Post Office called the route Foreign Air Mail Route No.2 (FAM2). In reality it was the *first* of many foreign air mail routes. The second route started fifteen days later between Miami and Havana and was designated as Foreign Air Mail Route No.4. Somewhere along the line the numbering did not reflect the order of coming into being. Foreign Air Mail Route No.1 (FAM1) between New York and Montreal did not commence until October 1, 1928.

The first contract was from October 15, 1920 to June 30, 1921. Hubbard's successful bid was as follows:

- to carry mail between Seattle and Victoria at not more than 600 pounds a trip.

- not to exceed an average of ten round trips a month at the rate of $200 a round trip.

The contract placed some interesting conditions on Eddie:

- he must carry the mail in a safe and secure manner, free from wet or other injury.

- not to commit the care or transportation to any person under sixteen years of age nor to any person undergoing a sentence of imprisonment at hard labor.

- he must post a $2,000 bond.

The Post Office wanted to expedite United States mail to and from the Far East. Mail for the Far East arrived in Seattle by train from the eastern states and often the train was late. As a result the mail missed a ship, which had just left Seattle, bound for the Orient via Victoria, B.C. This would mean the delay of a week or more until the next ship left Seattle. The President Line and various Maru ships leaving Seattle for the Far East, stopped at Victoria to pick up passengers, cargo and mail. With the start of flying mail to Victoria, the late mail from the East was able to catch ships in Victoria and not have to wait in Seattle until the next Orient bound liner.

Conversely, ships from the Far East stopped at Victoria to discharge passengers, cargo and mail. This included Canadian Pacific Empress Liners on their way to Vancouver, B.C. All United States mail carried by C.P. Empress ships went by train from Vancouver to Seattle, another day's delay. In many cases Seattle-bound ships did not arrive at their destination until a day or two after docking in Victoria. By flying the Orient mail to Seattle, importers received shipment invoices allowing them to contact their brokerage firm and obtain the necessary paperwork expediting clearance of their incoming cargo by one or two days. Silk shipments with a value of six million dollars were common. The record silk cargo was eleven million requiring 25 rail cars to take it to the eastern market. A saving of one or two days was worth a great deal of money to the importers. This air mail service was so important to importers it lasted until June 1937.

SEATTLE
VICTORIA
SEAPLANE
MAIL

Victoria Post Office Handstamp Used On Air Mail To Seattle

Seattle Post Office Handstamp used On Air Mail To Victoria

Eddie also flew local mail both ways between Seattle and Victoria and like the Far East mail there was no extra cost to the sender other than the existing surface postage rates. A few years later when air mail flights went from Seattle to other parts of the United States, American air mail stamps were required. To facilitate this, the Post Office in Victoria sold American air mail stamps. By 1927 the volume was forty dollars a month, mostly for letters to California and New York. The U.S. air mail rate at that time was ten cents for the first ounce. Both American and Canadian post offices had rubber handstamps which said, "Seattle-Victoria Seaplane Mail" or "Victoria, B.C. to Seattle, WA, Via Seaplane." All envelopes were stamped at both post offices and placed in a mail pouch either for Seattle or Victoria.

The first United States air service utilized by the Chinese Postal Service was the Victoria-Seattle air mail route. The Directorate General of Chinese posts ordered the Shanghai and Canton Post Offices to deliver 50 pounds of commercial mail each trip. Like mail from Canada, the Chinese mail did not require air mail stamps and there was no extra charge or rate for this service.

Finally the big day arrived and some 50 people stood around the ramp of Boeing's Lake Union hangar on a raw and windy October 15th. They had gathered to see Eddie Hubbard inaugurate the United States' first regularly scheduled international air mail service.

Seattle Postmaster Edgar Battle holding the first mail bag to start the air mail service to Victoria, B. C., October 15, 1920. As Hubbard lifted off Lake Union Battle remarked, "This is a day to remember."

Bill Boeing wishing his friend Eddie Hubbard a safe journey on his new international air mail venture. Eddie had the use of Boeing's own plane until he purchased the B-1 a few months later. Hubbard rejoined the Boeing Company in 1927.

Seattle Postmaster Edward McGrath proclaimed the event as historical and "A DAY TO REMEMBER." Hubbard and McGrath loaded five mail sacks aboard Bill Boeing's personal CL-4S seaplane; then, at 2:30 p.m. Eddie took off for Victoria making the 78-mile trip in one hour.

He reached Rithet's Outer Wharf, Victoria, in time to put the mail aboard the *Africa Maru,* which had left Seattle earlier in the morning bound for Japan. Eddie was met by R.B. Rithet, owner of the wharf, and Bill Boeing's Buttle Lake fishing partner. Also on hand was H.D.(Harry) Barnes, Rithet's Outer Wharf manager. To quote Harry Barnes Jr., "I remember when I was five or six dad would appear in the kitchen with wads of sweaters and let my older brother drive us down to the Outer Wharf. Eddie would come along shortly, the mail was loaded and away he would go. I always thought what marvelous fun it must have been." Barnes Jr. went on to say, "Rithet originally came from San Francisco and in Victoria had many holdings as well as being a shipping agent for the Blue Funnel Line and the Japanese Maru ships. He was an insurance broker, large grocery wholesaler and also owned an iron works."

Hubbard delivered United States mail for the Far East and Victoria to Rithet's Outer Wharf in Victoria. That was where the ships from the Far East docked. Eddie also picked up mail to the U.S. from the Far East and Victoria at the same location. The photograph was taken by Eddie as he approached for a landing.

Hubbard picked up the local mail from Victoria Postmaster Harry Bishop and Assistant Postmaster George Gardiner at a dock in front of the Empress Hotel. Because of the huge crowd that turned out to shake his hand, Eddie was only able to return to Seattle just before dark with Harry Barnes as his first passenger from Victoria.

Eddie receiving local mail for the United Sates from Victoria Postmaster Harry Bishop and Assistant Postmaster George Gardiner along with Jack Rithet. Eddie is climbing aboard the CL-4S Boeing seaplane on October 15, 1920 for the first flight to Seattle. In the background is Victoria's Causeway and the Empress Hotel. After this initial flight all mail was delivered to and picked up at Rithet's Outer Wharf.

Eddie used Bill Boeing's seaplane for the first few months. After that most of his mail deliveries and pickups were in his own Boeing B-1 flying boat. For pickups at Victoria, Post Office officials in Seattle were notified by wireless as to the time incoming steamers would be arriving at Victoria. Eddie and his flying boat would be dispatched. Hubbard made a go of the mail run from the start. He was a rare mixture of a man — a veteran pilot who knew the value of a dollar and recognized the opportunities opened up by the airplane. Most early fliers lacked this combination.

Cert. No. 108
File No. 553.1

THE AIR BOARD.
Ottawa ---------- Canada.

This certifies that -------- Edward Hubbard ---------------------------

whose address is ------- 56 Roanoke Street, SEATTLE, Washington, U.S.A.

and whose photograph is attached, being a citizen of the United States of America and having produced evidence that he is a qualified pilot, is authorized until the first day of July 1921 to fly in Canada as a Commercial pilot of light flying machines (hydroplanes), pursuant to the decision of the Air Board made the 17th day of May 1920, under the provisions of Para. 133 of the Air Regulations 1920.

This authorization is issued subject to compliance with all the terms of the said regulations and to cancellation at any time for cause. The holder of this certificate may not carry passengers or goods for hire between points in Canada.

Dated this ...26TH....day of ..NOVEMBER...... 1920.

For the Air Board.

Stanley Scott
Controller of Civil Aviation.

This Certificate is hereby extended for a further period of six months, viz. from July 1st 1921 to December 31st 1921.

[signature]
Acting Controller of Civil Aviation.

This Certificate is hereby extended for a further period of six months, viz. from January 1st 1922 to June 30 1922.

[signature]
Acting Controller of Civil Aviation.

Signature Edward Hubbard

(OVER)

Canadian Air Board certificate issued November 26, 1920 authorizing Eddie Hubbard to be the pilot of light flying hydroplane machines in Canada. However he is not to carry passengers or goods for hire between points in Canada.

UNITED STATES OF AMERICA
DEPARTMENT OF COMMERCE
AERONAUTICS BRANCH

PILOT'S IDENTIFICATION CARD

This Identification Card, issued on the 16th day of Dec., 1927, accompanies Pilot's License No. 1154

Age 38
Weight 145 Color hair Light
Height 5'8" Color eyes Brown

FORM R-11

Pilot's Signature.

UNITED STATES OF AMERICA
DEPARTMENT OF COMMERCE
AERONAUTICS BRANCH
FORM R-10

OFFICIAL NO. 1154

This Certifies, That Edward Hubbard,

whose photograph and signature accompany this license, is a TRANSPORT PILOT of civil aircraft of the United States and entitled to the privileges of all classes of licensed pilots.

This license expires June 15, 1928

Clarence M. Young
DIRECTOR OF AERONAUTICS

License Renewal

MAIL CARRIED BY EDDIE HUBBARD ON FOREIGN AIR MAIL ROUTE NO.2

This envelope was carried on the first flight from Seattle to Victoria at 2:30 p.m., October 15, 1920. The time on the postmark is interesting. It is neither a.m. nor p. m., but "12 M" which is noon. "M" is for meridian.

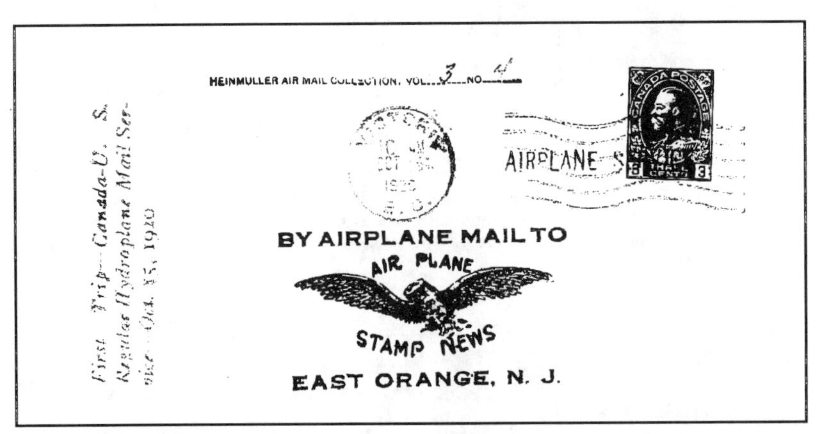

A cover flown from Victoria to Seattle on the first flight south leaving Victoria at 5:15 p.m. The Victoria Post Office had a handstamp, "Airplane Service," which is partially over the stamp. Both covers were prepared by an East Orange, New Jersey stamp dealer, A. C. Roessler. He also published a periodical, *Airplane Stamp News.*

Roessler is very well known to air mail cover collectors. If it were not for his efforts many first flight covers would not exist; however, there were a number of times he produced covers that were questionnable. More on that later.

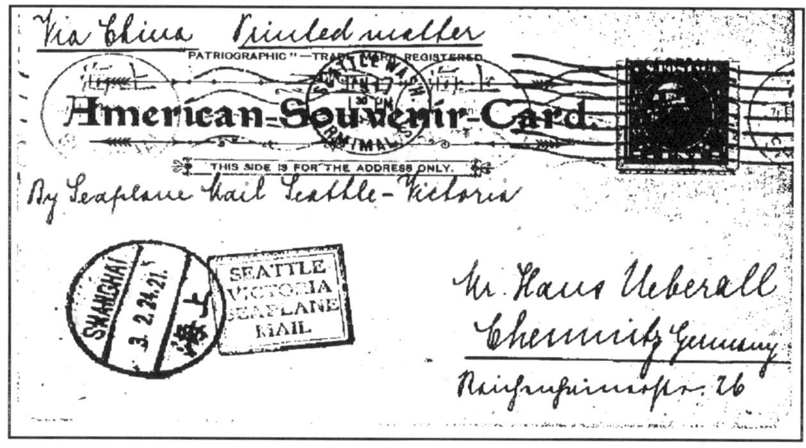

The Seattle Post Office used a handstamp to identify air mail for Victoria shown on the above post card to Germany via China.

The Victoria Post Office changed from the “Airplane Service” handstamp to the above round handstamp. This cover is addressed to A. C. Roe, which is a name A. C. Roessler used at times.

Roessler heard of a plane crash in Seattle on July 26, 1925, and assumed it was Hubbard. Gerald Smith, a friend of Hubbard's, had crashed without serious injury into Green Lake. Even if Eddie had crashed he always had the use of Bill Boeing's personal seaplane. Our friend, Roessler wrote "By *S. S. Pres. McKinley*" on the side of the envelope indicating it went by ship. As it so happened the *McKinley* was in the Orient at the time. Why would the Seattle Post Office use its handstamp if the cover did not go by plane? Note the oblong handstamp which Roessler used. It does not appear on non-Roessler covers such as the one below.

If there was an interruption in air mail service how did this non-Roessler cover to Japan fly with Eddie on August 7, 1925, when the top cover on this page went by "so-called ship" the day before. Note this cover does not have Roessler's oblong handstamp.

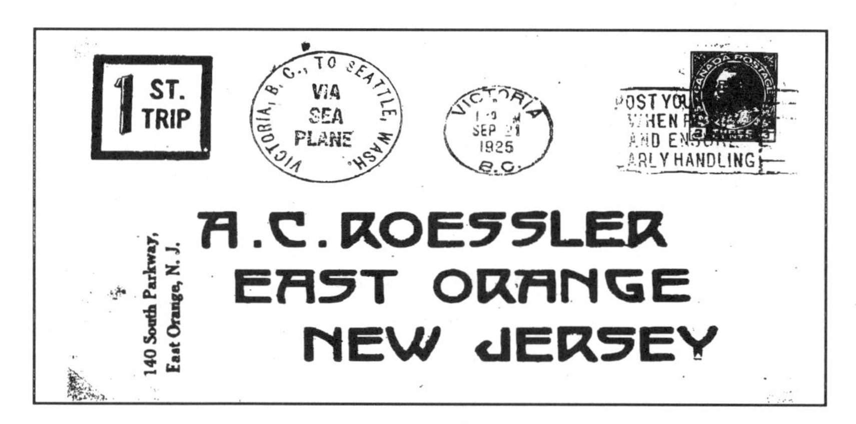

Roessler, with his unfounded claim, stated the interrupted air mail service commenced again between Victoria and Seattle, September 21, 1925, and produced covers with a red handstamp "1st Trip." He also said the interrupted service between Seattle and Victoria commenced again on November 11, 1925. Why would Hubbard only take mail one way? If Roessler was correct then logically when the air mail service came out of the so-called interruption it would be both ways. FAM2 is not the only air mail service on which Roessler or Roe produced dubious covers for sale. Recent research by aerophilatelists (1) has uncovered many figments of his imagination.

(1) One who collects and studies air mail postage stamps and all related material concerned with postal history which includes covers.

Cover produced by the Northwest Chapter of the American Air Mail Society for the society's annual convention in Seattle, May 24-26, 1991.

Every newspaper in Seattle and Victoria had front page headlines telling about this great international event. The second trip from Seattle was made the day after the inaugural flight carrying eighteen bags of mail and two passengers, the returning Harry Barnes and H.B. Davenport of the Blue Funnel Line. Both of these gentlemen were the first passengers from Seattle to Victoria. Heavy southeast winds made docking at the Outer Wharf a problem and Rithet immediately announced he would have a wharf built on the sheltered side for future mail deliveries. Eddie took time off to have lunch in Victoria with an old pal, George Weiler, with whom he had flown in San Diego as an instructor in WWI. The Weiler's were originally from San Francisco. George owned and operated Weiler's Auto Supply House on Douglas Street in Victoria. Mary Weiler, George's sister-in-law, said her husband Otto Jr. was a bachelor in those days and often took trips with Eddie to Seattle.

Eddie's flying instructor buddy, George Weiler, at San Diego in 1917. The Weilers moved to Victoria from San Francisco before the turn of the century. His father Otto founded a very successful furniture factory and retail store. George was in the automotive business, then moved to Sooke as manager of the water works.

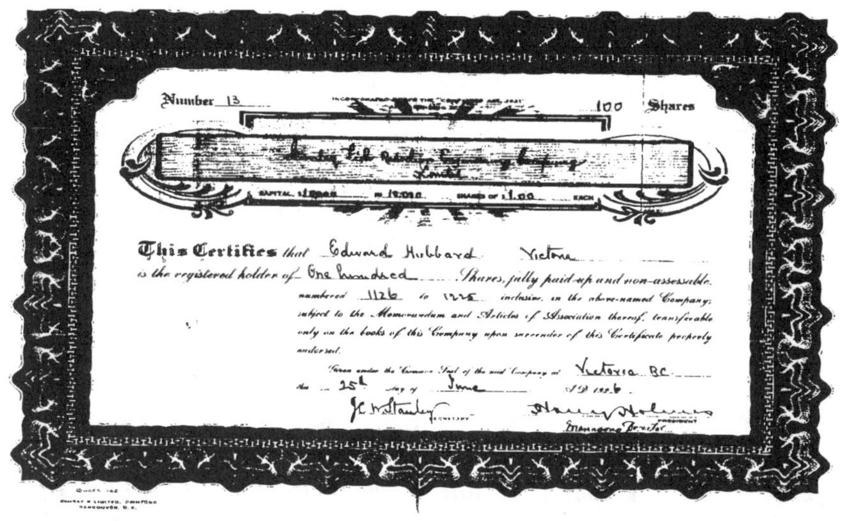
Number 13 100 Shares

This Certifies that Edward Hubbard Victoria
is the registered holder of One hundred Shares, fully paid-up and non-assessable.
numbered 1126 to 1225 inclusive, in the above-named Company;
subject to the Memorandum and Articles of Association thereof, transferable
only on the books of this Company upon surrender of this Certificate properly
endorsed.

Given under the Common Seal of the said Company at Victoria BC
the 25th day of June A.D. 1926.

Secretary

Managing Director

One of the few bad investments Eddie ever made was with George. Somehow they each put one hundred dollars into an outfit called Stanley Fish Reduction Engineering Company Limited in Victoria. Along with half the Seattle Metropolitans hockey team they each purchased 100 shares at $1 per share. By 1938 the company no longer existed.

In case you are ever on "Jeopardy" or need the answer for "Trivial Pursuit," the Seattle Metropolitans beat the Montreal Maroons at Seattle in 1917 and became the first American city to win the Stanley Cup, emblematic of world hockey supremacy.

The first few mail trips were not without incident. On the fifth trip Eddie was so late getting out of Victoria he had to fly most of the way in darkness. His trick on landing safely in the dark at Lake Union was to watch the lights on shore as a guide to his altitude. By doing this he was able to put the aircraft gently onto the water. On the next trip he had engine trouble and was forced to land at Port Ludlow on the Olympic Peninsula. In a short time the engine was once again acting normal and he was able to get airborne and head back to the hangar. A few days later on the trip north he ran into gale winds which made the flight almost twenty minutes longer than usual, using more fuel. Returning to Seattle in the dark, Eddie ran out of fuel and was forced to land at the mouth of Lake Washington Canal. After drifting for an hour and shouting for help he was able to attract attention from a homeowner on the beach. The local citizen rowed out with gas, enabling an exhausted pilot to finish the trip to the Lake Union hangar.

It was now time for Eddie to have his own mail plane and he chose the Boeing B-1 flying boat with which he was very familiar. In order to finance the purchase he formed the Hubbard Air Transport Company and for cash he gave notes to Phil Johnson, Boeing General Manager, and Ed Gott, Boeing Vice President, who became part owners in his company, signing an agreement on October 21, 1920. In a memorandum from Gott to Johnson, dated November 28, 1920, he mentions that Hubbard had paid the interest on his note but not any toward the principal. Gott assumed that Hubbard had not asked for an extension on the paper account because of other demands. Gott ends by saying, "I sincerely hope he is doing well and that the little part you and I played in the matter has turned out satisfactorily." Little did they realize the wealth Eddie Hubbard was about to accumulate down the road with investments in real estate, bank shares and Boeing stock.

Eddie's first mechanic, Les Hubbel, later said, "The B-1 was a good airplane with an excellent hull. Hubbard had very few navigating instruments, no weather reports, no radio beams, no method of communication at all. When the fog was heavy on the route to Victoria he often put the plane down on the water and taxied for miles. Unless the water conditions were bad you could taxi the plane at 40 miles an hour. You could sail it too. Hubbard had to land between two long piers at Victoria. The wind was usually behind him when he took the ship out to start his takeoff in open water. He used to walk back and forth on the wing, let the wind catch the tilted surface and sail the plane to open water! Then he would swing the propeller until the engine started, jump into the cockpit, and fly it away. The plane would carry a lot of weight for its size. Sometimes Hubbard would take-off with mail bags stuffed all around him in the forward cockpit. He charged $25 to take a passenger to or from Victoria and sometimes the passenger had to sit on the mail bags!"

Eddie and his B-1 flying boat became very popular with children in Victoria and Seattle as it was the only plane that made frequent trips to and from both cities. Van Peirson, a retired United States Postal employee reminisced. "We moved to Seattle in 1922 and Eddie Hubbard flew his plane off Lake Union just eighteen blocks from our house. As a twelve year old I was facinated by the mail plane and Eddie became one of our heroes. As I recall he left for Victoria about 2 p.m. or 2:30 p.m., returning about 5 p.m. I thought he had the best job in the Post Office! He also carried passengers both ways who were met at each end by Custom's people."

Seattle to Victoria Aerial Postman Is Proving High Ideal

Eddie Hubbard Is Out to Prove That Carriage of Aerial Mails Can Be Effected Regularly and Made Commercially Successful.

Over Million Letters Speed Through Air In Seattle-Victoria Flights

Eddie Hubbard Attained 100 Per Cent. Efficiency In Ten Months of Arduous Aerial Mail Service.

Daily Trans-Continental Aerial Mail Service Operating in States

Mail Planes of U. S. Postal Department Flying In 200-Mile Relays From New York to San Francisco In Average Time of 59 Hours.

Glimpses From Airway Between Victoria and Seattle on Mail Plane

Times Representative Is Guest of Aerial Postman On 3,000 Feet Level Across Straits of Juan de Fuca.

It was time for Eddie's first report card. *The Victoria Daily Times* had four separate articles in the September 3, 1921 edition about the "Aerial Postman." The first headline proclaimed, "AERIAL POST IS COMMERCIAL SUCCESS," and reported over one million letters had been flown safely between Seattle and Victoria. His efforts have convinced postal departments the world over the benefits of flying the mail. As a result of Eddie's efficiency he was awarded the contract for the second year. *The Victoria Daily Times* sent one of their reporters on a mail run from Victoria to Seattle and the write-up took a half page to describe the many sights along the way.

During this same interview Eddie made a profound statement: "*I want to prove that aeroplanes can carry mails in regular and efficient service and I want to show that carriage of mails by air can be made a commercial success.*"

As our story unfolds you will see that Edward Hubbard, in no uncertain terms, made his point.

In June 1921 Eddie was awarded the contract for the second year and the United States Post Office reported during his first year — October 15, 1920 to October 15, 1921 — 103 round trips were made and there had been no accidents and no defaulted trips. In January 1922 Eddie was visited by two Post Office officials to discuss his air mail route. The first was C.F. Egge, General Superintendent of Air Mail Service, who told Hubbard that the recent legislation pending by Congress to do away with operation of government air mail service had nothing to do with his foreign contracted air service. The government intended to get out of the air mail business and would call for bids from airplane manufacturers to look after the service. Egge talked about air mail from Seattle, Portland and Spokane. He also mentioned that mail from Seattle to San Francisco took thirty-three hours by train. If there was air mail it would take just eight hours.

The second visitor was Walter Riddell, General Superintendent of Railway Mail Service. He was in Seattle to observe the Post Office's facilities to handle the ever-increasing Orient mail leaving from the Seattle Post Office. Prior to the air service to Victoria the largest part of mail volume to the Far East went out of San Francisco. Hubbard's service had shown the postal authorities the shortest and fastest way for U.S. mail to get to and from the Orient was from Seattle. Riddell noted that eight of every ten letters to the Orient was now leaving by Seattle and the last mail shipment from one factory was valued at a quarter of a million dollars. Everything from postcards to farm implements were being shipped.

During this period, American customs officials were increasingly concerned about narcotic and liquor smuggling by hydroplanes. They asked Eddie to assist them in reporting any aircraft he came across while doing the air mail run to Victoria.

For the third straight year the Post Office renewed the contract with him for the air mail route to Victoria with one change — the number of round trips a month was increased from ten to twelve.

Eddie received his second report card. Seattle's George Williams, Superintendent of Mails, made the announcement that the Seattle - Victoria mail route in twenty-eight months had carried 50 tons of mail and travelled 35,200 miles in 220 round trips without a major mishap. With nothing but salt water and a few islands between Seattle and Victoria, there was no need for an airplane to climb after takeoff. The B-1 flew just a few feet, sometimes only inches, above the waves and in so doing discovered the fuel-saving and performance benefits of flying in "ground effect." This ability to fly at ceiling zero enabled the B-1 to get through many times in weather that made overland flying impossible.

The Seattle Times gave Eddie his third report card in a one page article entitled, "HOW SEATTLE'S AERIAL POSTMAN DELIVERS HIS MAIL," quoting W.C. Van Devoort, Superintendent of Railway Mail Service: "It is the most important and most satisfactory of all our aerial mail routes —— because of great time savings." *The Times* went on to say it is estimated millions of dollars worth of foreign commerce had accrued to American businessmen because of the service rendered by Eddie Hubbard, aerial postman. When the mail was extra large Hubbard would make two trips in a day. One amusing incident happened when a Seattle-bound passenger arrived in Victoria from Japan and gave a stamped letter to Eddie for his wife. She received the letter at their home in the University District early that afternoon and was able to meet her husband at the Seattle dock at two in the afternoon!

In June 1923 Eddie got the shock of his life after compliments from the Post Office for moving 50 tons of mail in 220 accident free trips. He lost the contract to Alaska Airways Company, owned by Silas Rich. Their pilot, Anscel Eckmann, was to handle the mail run with the company's Curtiss HS-2L World War II bomber flying boat. This was one of the twenty-five HS-2L flying boats built by Boeing after the war and it soon turned out the big seven-passenger Curtiss was no match in economy with Hubbard's B-1. Alaska Airways had added two cockpits. The larger engine in the HS-2L used a lot more fuel and the maintenance costs were much greater. It wasn't long into the contract when Alaska Airways subcontracted Eddie to do the mail run in his B-1.

On his last trip before losing the contract Eddie had an interesting experience. When he arrived in Victoria with the mail for the Makuru, he found that the ship had left for the Orient, so Eddie took off and caught the vessel at Race Rocks about twelve miles from Victoria in the Straits of Juan de Fuca. Victoria Postmaster Bishop wired the ship that Eddie was on his way. When the mail plane came in sight the Makuru stopped engines as Eddie landed beside it and transferred the mail bags.

Alaska Airways had big ideas announcing the purchase of a Junkers aircraft and obtaining a landing site on Lake Union. Contracts were to be let for the erection of hangars, a machine shop and a depot. They were also going to add six Aero Marine flying boats to their fleet. The plan was to start a passenger service to Tacoma, Everett, Victoria and Vancouver, B.C. Like so many dreams in the early days of aviation it remained a dream.

Because of Eddie's absence doing the Florida Navy testing for Boeing, Alaska Airways had another Seattle pilot, Eddie's old friend and fellow Boeing employee, Herb Munter, take the air mail to Victoria. On one trip Munter and his mechanic ran into a 65-mile per hour gale and were forced to land on the shores of Portland Island near Sidney, B.C. They managed to pull their seaplane up onto the shore and had to wait overnight before a boat could take Munter to Sidney then on to Victoria by car. The two were able to fly the plane back to Seattle two days later.

The Orient mail was growing to such a huge volume the Post Office authorized placing postal clerks on ships of the Admiral Line leaving Seattle. These men did the same work as mail clerks on trains, sorting and classifying mail for speedier delivery when they reached port. In the last year, mail to the Far East had increased by 104,770 bags on the Admiral ships alone.

Alaska Airways Company did not bid on the Seattle-Victoria air mail run starting July 1, 1924. With their uneconomical aircraft and having to subcontract to Eddie Hubbard the year was a losing proposition. Again Eddie Hubbard was the successful bidder.

During the month of August total silk shipments through the Port of Seattle amounted to $40,000,000. That is a lot of money today. Imagine what it was in 1923!

The importance to American importers of the U.S. air mail route from Victoria was noted in the Seattle press. A 10 million dollar silk shipment was headed east in two special trains for Chicago and New York. The high cost of insurance and the fluctuations in the price of silk dictated the speed at which the Oriental shipments were being handled. It is impossible to calculate the hundreds of thousands of dollars, maybe millions, the air mail service saved American business between 1920 and 1937. The route was discontinued in 1937 after the Pan American Airways' Martin M-130 'China Clipper' flying boat began serving the Pacific with mail in November 1935. The 'Philippine Clipper' and 'Hawaiian Clipper' were added and passengers were carried starting in October 1936. Early in 1937 Pan Am, added a Sikorsky S-42B flying boat, 'Hong Kong Clipper,' extending its route from Manila to Hong Kong. Commencing July 1940 Pan Am using Boeing 314 flying boats, inaugurated a service from San Francisco-Hawaii-Canton Island and New Caledonia.

For the fourth time, on July 1, 1925 Eddie was the successful bidder and retained the Seattle-Victoria air mail contract. The only change made was an increase per round trip from $200 to $250.

During October 1926 there were two mishaps on the Seattle-Victoria air mail run. Eddie had various pilots look after the mail while he was in California acting as the West Coast dealer for the Fokker Corporation. The first accident was by pilot Percy Barnes. Barnes got lost in the fog and had to land at Cadboro Bay. On the way to shore he hit some driftwood, putting a hole in the hull. Jane Galbraith, secretary to Boeing Presidents Phil Johnson and Claire Egtvedt, loved to fly, and often went along with some of the pilots using Boeing produced planes. Jane was on this flight with Percy and in her words, "My being aboard was sort of illegal. I'm not sure anyone at Boeing knew about this. I didn't tell them, as I remember!"

Jane Galbraith O'Neill, secretary to two Boeing Presidents, Phil Johnson and Claire Egtvedt. Here she is holding a model of the Boeing 40-B mail plane. Jane loved to fly and whenever she could took 'flights' on the Seattle air mail run and the testing of the 40-B. Her husband Ralph O'Neill started the 7,800-mile New York Rio and Buenos Aires Line which later merged with Pan American Airways.

A few years later Jane fell in love with Ralph O'Neill, who had been the Boeing Airplane Company agent in Latin America. O'Neill left Boeing, moved to New York and became President and general manager of the New York Rio & Buenos Aires Line (NYRBA). Jane also left Boeing, married Ralph and worked for his airline. This 7,800 mile route later was merged with Pan American Airways. The story is well told in Ralph O'Neill's book, *Dream of Eagles.*

By the way, the hole in Eddie's B-1 hull was patched in Victoria, allowing Percy and his "illegal passenger" to return to Seattle two days later. Jane's last comment, "We got back safely but his bride was furious."

Victoria Daily Times

WEATHER FORECAST

TIMES TELEPHONES

VICTORIA, B.C., TUESDAY, OCTOBER 26, 1920 — 16 PAGES

PRICE FIVE CENTS

MAIL PLANE CRASHES IN CITY

VERY SEVERE EARTH SHOCKS ARE RECORDED

Opinion of Observers is Centre of Disturbance East of Philippine Islands

Records Made by Instruments Here, in Eastern Canada, U.S. and Japan

WRECKAGE OF SEAPLANE WHICH CRASHED THIS MORNING AT RUPERT STREET WHEN PILOT GERALD SMITH LOST CONTROL

SEATTLE PILOT FALLS THREE HUNDRED FEET WITH TRIFLING INJURY

Gerald Smith, Substituting for Eddie Hubbard, Fights Desperate Battle for Control of Waterlogged Machine in Air, and Escapes From Crash With a Broken Ankle; Flying Boat Dashed to Fragments Against Senator R. F. Green's Residence Early To-day.

Crashing his plane against the side of the residence of Senator R. F. Green, 502 Rupert Street, this morning, Gerald Smith, carrying Seattle air mail, was dragged unconscious from the wreckage and now lies in St. Joseph's Hospital. Smith, who was substituting for Eddie Hubbard, has a broken ankle and gashes about the head and face, but since admission to the hospital has recovered consciousness. His injuries are not considered dangerous.

MRS. R. F. GREEN ESCAPES DEATH AS PLANE FALLS

The second incident happened when Gerald Smith went for the mail in an Aero Marine flying boat. He left the Inner Harbor at 9 a.m., heading for Seattle. Cec Ridout, as a young clerk with King Brothers Custom Brokers, located on the waterfront, watched Smith leave the harbor. "Just after take-off he experienced difficulty getting the tail up. It was all he could do to keep the nose from getting too high and incurring a stall." A few blocks from the Empress Hotel, Smith lost control and the flying boat flipped over, crashing into the front of Senator Green's house. The plane was a complete write-off and to the amazement of everyone, Smith pulled from the wreck by Senator Green and suffered only a broken ankle. He was rushed down the street to St. Joseph's Hospital for further observation. Cec heard about the crash, jumped on his bicycle and raced to the scene. The next day, after thanking all the people in Victoria for the flowers and gifts, Smith headed by boat to Seattle on a stretcher. The only explanation for the accident was that water must have gotten into the hull while he was picking up the mail. Senator and Mrs. Green were also most fortunate. They were both in the front room having morning tea when the plane hit the front of the house. Windows were shattered and flower stands knocked over, yet the Greens suffered no injuries.

Eddie had Gerald Smith fly to Victoria to make a mail pick-up on October 26, 1926. Smith used a friend's Aero Marine flying boat. After taking off from Victoria's Inner Harbor he could not gain altitude and after a few blocks crashed into Senator Green's house. Despite the total wreck of the aircraft Smith escaped with only a fractured ankle!

Pilot Smith said of the crash, "I nosed the plane out into a wide circle but ran into choppy water and decided to take off right away because my course was fouled by an incoming boat. I remember making a second circuit of the harbor, as the machine was tail-heavy and I had all I could do to hold her down. There may have been water in the fuselage, though I don't know for sure. At any rate she was down by the tail and I was trying to keep her from nosing up into a stall. The strain on my arms from the pressure was growing too much for me and I think that the controls must have parted. At any rate the machine got out of control. It did a half-roll onto its back and still I thought I could bring it out all right. After that I don't know what happened. I crashed."

With prohibition at its peak, bringing liquor from Canada into the U.S. became very profitable pastime. A rumrunning flying boat was captured at Lake Washington by Government agents. As previously mentioned, Eddie had been asked by authorities, when on his mail run, to be on the lookout for strange planes which might be carrying illicit liquor. The pilot and owner of the captured flying boat said he recently bought it from Eddie Hubbard and had installed a new engine. It was reported the Curtiss flying boat was worth $10,000 and Hubbard paid the Government $800. No wonder he died a millionaire!

Boeing B-1E flying boat used by Vern Gorst and Percy Barnes after Hubbard gave up the contract on June 30, 1927. In 1929 they became 'Seattle-Victoria Air Mail Inc.' Barnes was the pilot. This air mail service was discontinued by the United States Post Office on June 30, 1937.

It was time for Eddie to give up the air mail route from Seattle to Victoria. He did not bid on the contract commencing July 1, 1927. The successful bidder was Northwest Air Services Inc., whose President William Strain also was head of Sunbeam Coal Mines at Renton. Northwest Air Services had purchased the B-1 from Hubbard and operated it out of the Renton airfield until June 1928. Starting July 1,1928 Vern Gorst and Percy Barnes were the successful bidders and operated for one year under their own names, also from the Renton airport. They purchased a Boeing B-1E which was larger than the B-1 and had an enclosed cockpit. With its bright red hull it soon became known as the "red boat." In 1933 they formed Seattle-Victoria Air Mail Inc., and carried the mail until the Post Office discontinued the service on June 30, 1937.

During the six years and nine months Eddie flew mail from Seattle to Victoria and back he travelled 150,000 miles and wore out six Liberty engines.

ALL THREE BOEING AIRCRAFT FLEW THE MAIL ON UNITED STATES FOREIGN AIR MAIL ROUTE No 2 BETWEEN SEATTLE, WA. AND VICTORIA, B.C. FROM OCTOBER 15, 1920 TO JUNE 30,1937.

The "gang" at the Lake Union Hangar with Eddie Hubbard in the cockpit of Bill Boeing's CL-4S seaplane. This aircraft was similar to fifty trainers produced for the United Sates Navy by the Boeing Airplane Company.

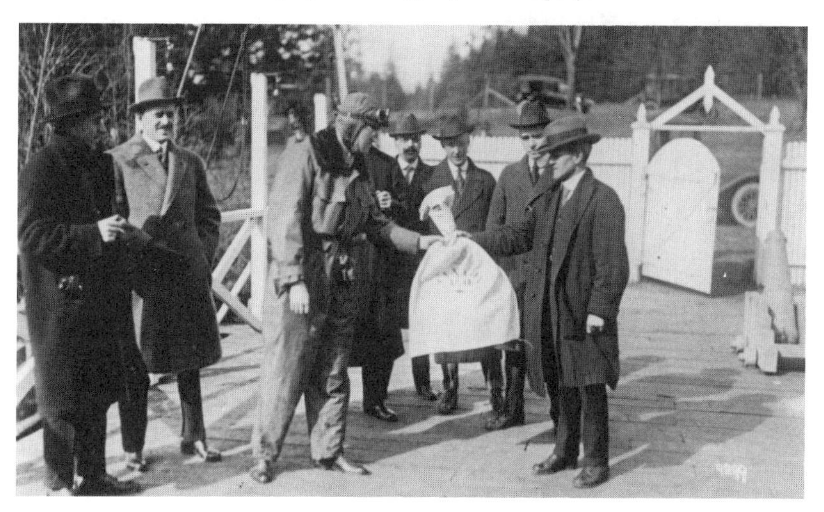

E. S Knowlton, presenting a Canadian air mail bag to William E. Boeing for the return flight from Vancouver, British Columbia, to Seattle, Washington, March 3, 1919. This became the first North American international air mail.

William Boeing and his personal seaplane at Coupeville, Whidbey Island. It likely took place on the Sabbath, as everyone appears to be in their Sunday best.

Seattle Postmaster Edgar Battle handing Eddie Hubbard the first air mail from Seattle, Washington, to Victoria, British Columbia, October 15, 1920. This was the start of United States Foreign Air Mail Route No. 2 which lasted until June 30, 1937.

October 15, 1920, Eddie Hubbard, in front of the Empress Hotel, Victoria, B. C., picking up the first flight mail for Seattle, Washington. Next to Eddie are Victoria Assistant Postmaster George Gardiner and Postmaster Harry Bishop.

One of twenty-five Boeing Model 40-B mail and passenger planes built for the San Francisco-Chicago United States Post Office Contract Air Mail Route No. 18.

Eddie Hubbard and passenger at the Lake Union hangar.

Custom Boeing Model 80 manufactured for the Standard Oil Company of California -- now Chevron.

First Model 40-B mail planes off the assembly line receiving last minute tuning at Sand Point air field, Seattle, June 14, 1927. Before the 2,000-mile air mail route from San Francisco to Chicago commenced July 1st, Boeing Air Transport distributed twenty-four of these aircraft along the route. Each plane flew in approximately 500-mile hops.

A World War I surplus Standard J seaplane on Boeing "C" type pontoons. This aircraft was used as a backup to Eddie Hubbard's flying boat on the Seatttle-Victoria mail run when Bill Boeing's personal seaplane, the CL-4S, was not available. United States mail for the Far East is being unloaded in Victoria, B. C., to be put aboard a ship leaving for the Orient.

Eddie Hubbard leaving Lake Union with a passenger in the "Aerial Taxi." Hubbard and Bill Boeing formed the "Aerial Taxi" business in 1919. When Hubbard left the Boeing Company in 1920 for the air mail contract from Seattle to Victoria they still carried on in the air taxi business.

Refueling at Anacortes, Washington, February 27, 1919 on second attempt to reach Vancouver, B. C. Facilities owned by Standard Oil Company of California. Aircraft owners preferred Standard's Red Crown gasoline. At the time it was the highest octane available on the West Coast. Charles Lindbergh used Red Crown for his famous trans-Atlantic flight.

Eddie Hubbard checking the camera on a 1921 Vancouver Island vacation with his wife Mildred. They took four Victoria friends along in Bill Boeing's personal seaplane.

Canadian Customs officials clearing the first foreign aircraft to land at Cadboro Bay, Greater Victoria, B. C., June 12, 1919. Bill Boeing was the passenger, flown by his test pilot Eddie Hubbard. Boeing gave an address to the Victoria Rotary Club at the Empress Hotel.

The B-1 after Hubbard's crash at Victoria on March 29, 1923. The aircraft was taken back to the Boeing plant on the Duwanish River and rebuilt in 57 days.

The second crash on the Seattle-Victoria mail route occurred on October 26, 1926. Hubbard had sent Gerald Smith to pick up the Victoria mail. Shortly after take-off Smith crashed into the home of Senator Green. Smith's aircraft was a complete write-off. Fortunately he escaped with only a fractured ankle.

Eddie Hubbard on the hangar ramp, Lake Union, outfitted in his flying suit.

Eddie Hubbard's B-1 flying boat on Lake Union, Seattle, in the days before smoke pollution was frowned upon.

Chapter 14
Interesting Passengers

BRITISH AVIATOR TELLS OF FLIGHT

Navigator With Alcook in Trip Across Atlantic Is Guest of Aero Club.

AIR TRANSPORT REALITY.

Sir Arthur Whitten-Brown Declares New Era of Service Has Already Arrived.

FAMOUS AVIATOR REACHES SEATTLE

Sir Arthur Whitten-Brown, One of Transatlantic Flyers, Will Lecture Tonight.

SEA. TIMES DEC 4, 1919

MAKES FLIGHT OVER CITY

Three months after Eddie and Bill Boeing made their historic air mail flight from Vancouver, B.C. to Seattle another pathfinder visited Seattle. Arthur (Teddy) Whitten-Brown, as navigator, with pilot John Alcock made the first non-stop crossing of the North Atlantic on June 14, 1919 from Newfoundland to Ireland, in a Vickers Vimy biplane. For this outstanding achievement both aviators were knighted and received the *London Daily Mail* prize of $50,000. Unfortunately Sir John Alcock was killed when he crashed in bad weather over France fifteen days after Sir Arthur arrived in Seattle.

Brown's visit was sponsored by the Aero Club of the Northwest to give a speech with slides at the Masonic Temple. Prior to his talk, before a sell-out crowd of aviation enthusiasts, he was the honored guest at a dinner held at the Army and Navy Club.

Earlier in the day Sir Arthur was given a tour of the city by Eddie Hubbard in the latest aircraft built by the Boeing Airplane Company, the C-700 (Bill Boeing's personal plane was later modified and designated CL-4S). It was so new that Hubbard had taken it up only twice on test flights. Eddie's job that day was to distribute leaflets over the town advertising the Red Cross Christmas Seal Campaign and to tell everyone about the world renowned navigator's lecture.

That evening Brown captivated his audience, opening with, "The age of air transport is not coming, it is here." Seattle's landing field problems are noted in an earlier chapter; however, Sir Arthur dwelt on this problem by pointing out the necessity of, as he put it, "Proper landing ports if commercial aviation is to become widespread at once."

Brud Nute, whose father lost Eddie to Bill Boeing, relates an interesting story about the air mail pilot. Brud's mother had passed away and his father decided it was a boy's private boarding school in Victoria for his son. "I had a tooth problem and had to go home to the family dentist in Seattle. I was due to return on the C.P.R. ship *Prince George* and somehow missed the sailing. My father immediately thought of Eddie Hubbard. He called Eddie and asked if he was going to Victoria that day and if so did he have room for me. Eddie apparently was flying to Victoria and I was taken down to the hangar on Lake Union. They put me in the front cockpit and we slid down the ramp into the lake. I will never forget the stench on that flight. In those days the engine was lubricated with castor oil and did it smell! We took off and did the whole trip in one hour. On the way up we passed the *Prince George*, Eddie tapped me on the shoulder and pointed down to the ship I was supposed to be on. He descended and I waved as we flew past the ship. We landed in the harbor right in front of the Empress Hotel. I was put ashore and took the street car to the end of the line at Mount Tolmie heading for St. Michaels University boarding school. From then on I owned that school. I had flown in an airplane! When I say "owned the school" just picture the youngest kid there in a school of less than a hundred students. I was either nine or ten but as I say I owned that school for quite awhile. Sort of an afterthought but knowing my father he probably didn't pay Eddie anything for that trip. He relied on old friendships and felt that was one way of getting some of his money back from the best mechanic he ever had."

It seems that more than one Seattle boy attended private school on Vancouver Island. Townsend Jones Moore from Gig Harbor went to a school at Shawnigan Lake, north of Victoria, in the early twenties and often took a ride from Seattle to Victoria atop the mail bags Eddie was carrying. Moore later became Vice President of H.P. Pratt & Co., Seattle bond dealers. His son Charles still resides in the Sound city.

Meeker Gets Big Thrill on First Voyage in Airplane

Pioneer Devotes Chapter in New Book to Telling of Experiences in Flying Trip to Victoria.

NONAGENARIAN FLIES TO VICTORIA

Ezra Meeker, Pioneer of Oregon Trail, Arrives

Undaunted Veteran Will Return With Hubbard

Aged Pioneer a Cloud Fan

Takes Air Mail to Victoria

Flying Beats Steamboating

Pioneer Washington resident Ezra Meeker about to depart the Lake Union hangar with Eddie Hubbard on a flight to Victoria, B. C. to present a manuscript of his early days in the Pacific Northwest to the British Columbia Provincial Library. This was on his 90th birthday! Ezra was the founder of the town of Puyallup, Washington.

In May 1921 Eddie had another passenger on one of his air mail trips to Victoria. On his ninetieth birthday Ezra Meeker, famous pioneer of the Oregon Trail and founder of the town of Puyallup, flew for the first time and compared the fast trip to his first visit. Sixty-three years previously Meeker sailed from Seattle to Victoria on the steamboat *Constitution,* a trip which took seventeen hours. With Eddie Hubbard, Meeker, with white hair hidden beneath a leather helmet, cut that time to forty minutes.

Meeker made the trip to Victoria to visit old friends and to present the British Columbia Government Library with a copy of a recently discovered old manuscript about his early days in the Northwest. He had copied it by hand for the library. While this was going on, Hubbard flew back to Seattle to pick up more mail for the Empress of Russia. Late in the day Eddie returned his "young" passenger to Seattle.

George Williams, Seattle Post Office Superintendent of Mails, ready to take a mail run to Victoria with Eddie.

The next dignitary to be flown by Eddie was George Williams, Superintendent of Mails for the Seattle Post Office. He had the honor of being the first United States Postal Official to cross the international boundary by airplane while on official business. Williams was checking out the mail route on one of Hubbard's trips.

First Aerial Tour of Island Begins Here

On June 20, 1921 Eddie tied the knot and took his bride, the ex-Miss Mildred McClure, on a wedding trip to Vancouver, British Columbia, not by airplane as everyone might expect but by train. One month later, in Bill Boeing's seaplane, he flew his new bride to Victoria to pick up George and Sis Weiler. George and Eddie were pals from WWI flying instructor days in San Diego. Another couple from Victoria also joined the group, Inez Gonnason and Evan Hanbury.

Inez was a member of a well-known Victoria lumber family and Evan was for many years the McGavin Bakery manager in Victoria. Inez later became Mrs. Hanbury.

This was the first aerial vacation on Vancouver Island and all six climbed aboard and somehow squeezed into the Boeing CL-4S seaplane's two cockpits. The plane left Victoria's Inner Harbor on its way to Buttle Lake. Eddie was familiar with that area, as he flew Bill Boeing's fishing party from Campbell River to the lake a year before. In order to take time off for a vacation Eddie noticed a five-day gap in ship arrivals, giving him the green light for a well-earned rest.

Eddie and Mildred Hubbard in July 1921 on a Vancouver Island vacation using Bill Boeing's seaplane with friends George and Sis Weiler, Inez Gonnason and Evan Hanbury. All somehow squeezed into the plane for a trip to Buttle Lake near Campbell River and Great Central Lake near Port Alberni. On the way home Eddie treated the group to an aerial view of British Columbia's Gulf Islands.

Sis Weiler, Eddie and Mildred in roaring 20's bathing suits.

The party arrived at Buttle Lake two hours later. Camping, fishing and swimming occupied their time in beautiful July weather. After Buttle Lake they flew to Great Central Lake near Port Alberni on the western side of Vancouver Island. Fortunately Inez was a camera buff and left a photograph album of the event with her family of this exciting 1921 flying tour of Vancouver Island. On the way back from Great Central Eddie gave his friends an air tour of British Columbia's Gulf Islands, an extension of the Pacific Northwest's San Juan Islands.

Eddie and his wife Mildred on a British Columbia vacation at Buttle Lake near Campbell River.

Around the world in eighty days went out the window in 1926 when two men, Linton Wells, a newspaper correspondent, and Elwood Evans, a Detroit millionaire, circled the globe in 28 days, 14 hours and 36 minutes. Eddie helped them establish the new record by picking them up in Victoria as their ship, the Canadian Pacific liner *Empress of Asia*, entered the harbor. A speedboat carried the two 'record-breakers' from the ship to Eddie's flying boat, bobbing on the waves. It was 5:40 in the morning and 65 minutes later Hubbard dropped down onto Lake Washington at the Sand Point Naval Air station. Sailors carried the two men on their backs to shore. They were driven a short distance to the landing field where Army pilots flew them to Pasco, Washington. At Pasco another Army plane took over and a series of flights took them to their starting point, New York City, where the two claimed the world record.

Chapter 15
Hunting Accident

HUBBARD'S PLANE FALLS!

HOME EDITION

ALL THE NEWS THAT'S FIT TO PRINT

The Seattle Daily Times

THE ONLY SEATTLE PAPER OWNED AND OPERATED BY SEATTLE MEN

The next event in Eddie's life was tragic. He and a close personal friend well-known Seattle lumberman, C.W. (Ban) Bandy took off on a duck hunting trip to Sequim on the Olympic Peninsula in November 1921. Flying back, at 5 p.m., the hunters ran into a blinding snowstorm off Bainbridge Island. Visibility became zero and our pilot had no alternative but to land on the rough ocean.

The seaplane hit large waves and started to break up. Both climbed as high on the wreckage as they could to protect themselves from the waves. They had been holding onto the wreckage for over an hour when Bandy suggested he take the pontoon that had become free and try to get to shore for help. They saw a light, which was the Port Madison dock, and with the wind blowing that way they both figured the pontoon would carry Bandy to shore. Hubbard floated with the wreck wondering why Bandy hadn't returned with help. At four in the morning after being soaked and freezing cold for eleven hours the plane struck bottom, so Eddie knew he was close to shore.

He jumped into the water and with a few strokes was on shore and saw a number of summer houses but couldn't find a phone. Finally he came to an occupied house owned by Norman Waterhouse and was able to phone for assistance. Waterhouse and Hubbard aroused Capt. George Smith, master of the yacht *Olive*, and cruised along the shore searching for Bandy. They found the pontoon but not Eddie's friend.

Two days later Bandy's body was found by two teenagers not far from the pontoon and wrecked seaplane. The plane Eddie used for their hunting trip was not the one he used on the mail run. It was a reserve plane that now became a write-off.

Chapter 16
Bank Robbers and Bloodhounds

TRAP BANDITS

WEATHER

The Seattle Star

HOME EDITION

During the 20's almost once or twice a week newspaper headlines would report another bank robbery. Many were violent, ending in a shooting of either bank employees or bandits — sometimes both.

One sensational night break-in occurred at the Sequim State Bank on the Olympic Peninsula in March 1922. The robbers got away with $20,000 in cash plus the contents from many local citizen's safety deposit boxes.

When the bank was opened the next morning the robbery was discovered and the alarm was given to local police officers under the direction of Clallam County Sheriff Billy Nelson. Deputy Sheriffs Rex McInnes and Joe Priest were sent to capture the break-in artists.

The manhunt moved to Discovery Bay with more area lawmen recruited for the capture of the two bandits. The United States Coast Guard gave a hand in patrolling the strait in case of an attempted escape by boat. The two deputies calculated that the robbers would try to reach the Discovery Bay Logging Company's railroad trestle in order to make their escape. Their hunch was right and McInnes, at one end of the trestle, saw a man near the beach and told him to halt and was immediately hit by a bullet. The shot came from under the railroad bridge hitting the deputy in the hip. His partner, Priest, helped him to the Maynard Hotel.

Later that night Port Angeles Police Chief Van Welch and Jefferson County Sheriff Chase arrived with a posse to hunt the bank robbers. Also that evening a poker game was in progress at a recreation hall seven miles from Discovery Bay. The game was interrupted by two bandits wearing masks and sidearms. After they scooped up all the poker stakes and took everyones wallet pushing and shoving began. One player was shot in the back while holding his hands high in the air. Several hours later the gambler was dead.

It was assumed the two killers were also the bank robbers. Local lawmen sought outside help and phoned Seattle Sheriff Matt Starwich to bring his two bloodhounds to the area. Starwich was very well known in the State of Washington for bringing many misdeeds to a successful conclusion.

Starwich knew he had to act fast before the trail became cold. He immediately thought of Eddie Hubbard and his mail flying boat. Starwich gave Eddie an early morning call and took his bloodhounds to the Lake Union hangar. The bloodhounds, pilot and sheriff were airborne by 5 a.m., winging their way to sniff out the killers. Eddie had the sheriff and his eager dogs at Discovery Bay in fifty minutes to pick up the trail.

The bank robbers were captured and a check of their guns indicated they had shot and wounded Deputy McInnes but a different gun killed the poker player. Both robbers denied holding up the poker game and killing anyone. By now it was evident that the lawmen had two separate crimes on their hands. A few days later the two killers were captured and one received a life sentence.

Chapter 17
Victoria B-1 Crash

Headlines screamed across the front pages of Seattle and Victoria newspapers:

HUBBARD HURT
FALLING PLANE

THE ONLY SEATTLE PAPER OWNED AND OPERATED BY SEATTLE MEN

6 | FINAL EDITION Complete

The Seattle Daily Times

EXTRA

Victoria Daily Times

PRICE FIVE CENTS

SEATTLE FLYING BOAT CRASHES IN THE STRAIT

TO TAKE "DRIFT" OUT OF EDUCATION

On March 29, 1923 Eddie flew from Seattle to pick up Far East mail that had just arrived in Victoria on the *President Grant.* After take-off at Brochie Ledge, a navigation beacon, about a mile from the loading ramp at Rithet's Outer Wharf the B-1 suddenly slammed into the ocean. In Eddie's own words this is what happened:

> "For some unknown reason the wire which controls the rudder of my plane snapped just after I was taking off. I am at a loss to explain why the wire broke. I suppose it must have worn thin as a result of friction over

a long period. Once the wire snapped there was no way of controlling the rudder and the plane was quite unmanageable. As soon as I realized that my rudder was gone, I tried to make as good a landing as possible, and as I was flying only a few feet from the water when I hit the surface I was not at all hurt. When I hit the water I tried to untie the strap which held me to my seat, but my glove sticking to my hand prevented me from getting myself clear. Tied to the plane, I naturally went under the water with her and I couldn't get loose. I was not hurt by the fall but the water soon overcame me."

The accident had been witnessed by two commercial fishermen in the immediate vicinity and within a few minutes of the plane striking the water were alongside and they were able to release Eddie from the smashed plane and get his unconscious body into their boat.

The accident had also been seen by people watching its flight from the Dallas Road seafront, from the decks of the *President Grant* and by Tom Colley of the B.C. Pilotage Association. Colley and George Weiler, Eddie's flying instructor friend from WWI, sped to the scene. Efforts were made to revive the airman as the pilot boat headed for Ogden Point pier. At the dock a member of the *President Grant* crew gave Eddie artificial respiration and within several minutes he regained consciousness. In the meantime the ship's doctor arrived and ordered Eddie to be carried aboard the ship and put to bed. When the *President Grant* arrived in Seattle that evening Eddie was able to walk ashore!

Not a man to waste time Eddie took the C.P.R. morning boat from Seattle to Victoria to inspect his wrecked flying boat. He then made arrangements for it to be transported to Boeing in Seattle.

Hubbard's B-1 flying boat at Victoria Machinery Depot wharf after crash near Victoria on March 29, 1923. The B-1 was returned to the Boeing plant in Seattle and rebuilt in 57 days!

Shortly after the crash two bags of mail were recovered at the scene of the accident and the crumpled flying boat was towed to the Inner Harbor by tug and lifted out of the water at the Victoria Machinery Depot where the remaining nine bags of mail were recovered. The B-1 was taken to the Boeing Airplane Company in Seattle, rebuilt, and was back in service by May 25th. Eddie continued to use this "one of a kind" flying boat on the mail run until he went back to work for Boeing in 1927.

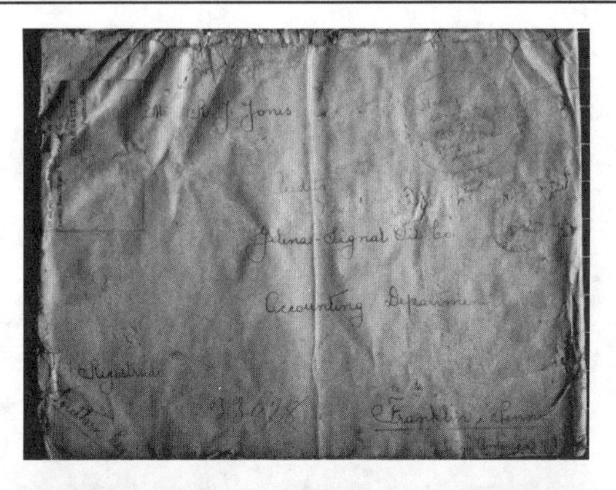

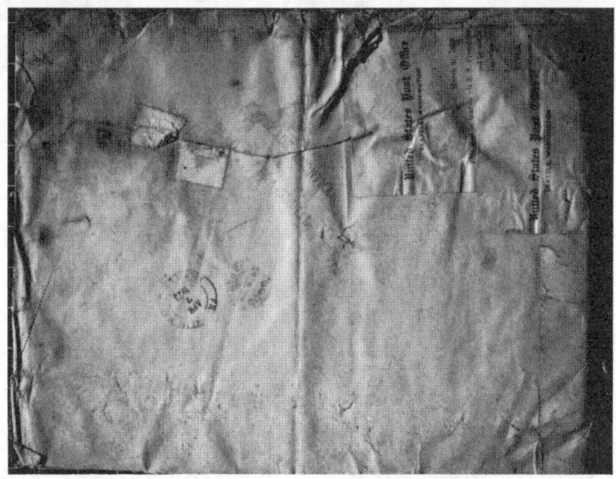

An envelope recovered from the Victoria B-1 crash addressed to the Auditor Galena-Signal Oil Co., Franklin, Pennsylvania, from their office in Yokohama, Japan. The stamps were washed off by the sea water. The United States Post Office in Seattle placed a printed sticker on the envelope, "March 31, 1923 this mail in transit via S. S. President Grant from Yokohama was damaged by water in an accident to the seaplane in service between Seattle and Victoria, B. C., Edgar Battle Postmaster." The letter left Yokohama March 8, arrived Victoria March 29 and was delivered to Pennsylvania April 7.

Chapter 18
Demonstrating Navy Trainers

By October 1923 Eddie, as well as flying mail for Alaska Airways, was back as a test pilot for Boeing. The first new trainer was the NB-1 (Model 21) and was aimed at the Navy. After testing it Hubbard declared it the best plane ever. After much debate between Claire Egtvedt, Boeing's Chief Engineer, and Edgar Gott, Boeing President, as to whether Hubbard should go to the Naval Air Station at Pensacola to demonstrate the NB-1 they finally agreed to have him put on a spin test for the navy. The reason for the debate as to whether Eddie should fly the plane might be the possible resentment from Navy pilots if a civilian showed the plane's ability. Hubbard had a lot of contact with Navy pilots and they believed he could present the new plane in a way that would cause no resentment. Boeing was in competition with three other builders.

After several days of testing, Lieutenant G.L. Compo, the engineering officer on the trial board, came to Hubbard. "We can't get your plane into a spin, it's too stable."

"Isn't that good?" replied Eddie.

"As long as other planes spin, we can't buy a trainer that you can't learn to spin in."

Hubbard decided they could make the plane spin by adjusting the stabilizer to its maximum up angle. The technique of getting into a spin was to stall a plane by slowing it to less than flying speed, then kick the rudder over, which should put it into a spiral dive. The stabilizer adjustment would cause it to stall, or lose flying speed at a steeper nose-up angle, and should ensure that it would go into a spin.

Hubbard took Compo along to try it. They went up to 2,600 feet. Hubbard stalled and kicked the rudder hard over. They went into a normal spin but after about four turns Hubbard noticed the spin was turning faster and getting flatter. The nose wouldn't stay down. He kicked the opposite rudder and pushed the stick over. The controls had no effect. The earth was twisting below so he gunned the motor to put a slipstream on the tail to improve control. It only spun more rapidly. Eddie reasoned it must be turning so fast that the slipstream misses the tail. Then it occurred to him to readjust the stabilizer. Drawing up all his strength he tried to move the stick but the load on the stabilizer was too great. He couldn't move it.

Pensacola Bay, gyrating wildly, was coming up fast. All he could do was pull back on the elevators to make the plane flatter, and check the drop as much as possible. They kept whirling flatly and then hit. One pontoon twisted off under them and there they sat on the water. Hubbard and Compo looked at each other blankly for a moment, then climbed up on top of the upper wing. They were talking it over when Claire Egtvedt, arrived full speed across the bay.

"What did it look like?" Hubbard asked Egtvedt.

"Like a falling maple seed. A flat spin."

"Whoever heard of a flat spin?" said Hubbard. "There is no such thing."

Compo began to laugh. "But we did it. Safest way I ever saw to wreck an airplane. You come down so easy."

Egtvedt wasn't so sure he was being funny.

"It's nothing," said Compo. "I want to be the first to go up again when you get it fixed."

They hashed and rehashed the accident. While they were having a Curtiss pontoon installed, they lengthened the rudder, moved the engine forward two inches, and put the stabilizer back to zero. "You'd better wait till we try it," Egtvedt told Compo. In the new condition, they found they could successfully tailspin it and bring it out without going into the uncontrollable flat spin.

In the spring of 1924 the Navy declared Boeing the winner of its competition and ordered forty-nine of the new trainers. Hubbard's luck was still riding with him. Testing a new Boeing plane for the Navy he fell over three-quarters of a mile into Lake Washington. The aircraft was a total wreck but Eddie came out miraculously with only a fractured wrist and some bruises. Spectators on the shore assumed the pilot would be a dead man.

Chapter 19
San Francisco-Chicago Mail Contract

Phil Johnson, Boeing General Manager, in 1927 announced a new type of air mail plane that would be ready for trials in two weeks. This was the trimotor Model 80. On the other hand Edgar Gott, former Boeing President, now working on his own, was collaborating with Eddie Hubbard and Donald Douglas, designer of the Douglas Cruisers used on the Army's around the world flight. Their company was called the Seattle Airplane Construction Company and they were negotiating with the postal authorities for a north-south air route from Pasco, Washington to Las Vegas, connecting with transcontinental air mail at Elko, Nevada. If they were successful they would not be using Boeing planes but would use a new type of Douglas plane. Hubbard suggested to postal officials that if the aircraft had additional capacity on a trip they be allowed to also carry regular 2c mail. The authorities didn't approve of this idea which may be the reason the consortium dissolved.

Delay after delay occurred with bids and awarding the Pasco-Las Vegas air mail route. Finally a bid was received from Walter Varney for the route. Also a bid came in from Vern Gorst for the Seattle-Los Angeles route.

In the meantime Eddie was testing the modified Model 40-B and was very enthusiastic over its performance. He then flew Boeing's new air mail plane, the 40-B, on a trial run to San Francisco to show the Postal Authorities what it could do. He left from the new airfield at Sand Point on Lake Washington.

Henry Ford was also building planes to carry the mail between Detroit and Chicago. Ford had purchased the Stout Metal Airplane Company and made it a division of the Ford Company. The eventual result was the Ford Trimotor.

Vern Gorst won the first Pacific Coast air mail service between Seattle and Los Angeles. Eddie Hubbard was associated with Gorst's new company, Pacific Air Transport, as was Claude Ryan who was operating an air mail route between Los Angeles and San Diego. Ryan was also in the aircraft manufacturing business. It was a Ryan plane that Lindbergh flew across the Atlantic. In a speech to the Seattle Chamber of Commerce, Gorst promoted an airfield at Georgetown in south Seattle, saying it would cut the time for the mail run, which was now seven hours between Seattle and San Francisco using Ryan monoplanes. The main reason Gorst was promoting a new field was because the County was about to lease Sand Point to the Navy.

Some big changes were taking place in the Boeing Airplane Company. Ed Gott, general manager, resigned after a dispute with Bill Boeing. Phil Johnson was named President and Claire Egtvedt became Vice President and chief engineer. Boeing himself assumed the chairmanship

For the past year and a half Eddie had been patiently waiting for the right air mail opportunity. He passed on the Pasco-Elko route and backed out of Gorst and his Pacific Air Transport Company's Seattle-Los Angeles run. For Eddie the big one finally arrived. The Post Office wanted completely out of the air mail transportation business. It was to be contracted out. In the meantime Phil Johnson thought about starting a company to carry passengers between Seattle, Vancouver and Victoria in order to demonstrate commercial aircraft. Claire Egtvedt, new Boeing General Manager, liked the idea and wished Eddie was around. Out of the blue Eddie returned to Seattle but showed no interest in a Puget Sound airline. The Post Office was about to put the New York-San Francisco route up for bid in two sections, New York-Chicago and Chicago-San Francisco.

"This is the opportunity of a century, Claire," Eddie said. "I've got all the figures on mileage and pounds of mail carried. If you can produce some mail planes I know we can operate them profitably."

Claire shot back, "Night flights? Are the beacons in all the way?"

"Every twenty-five miles."

"We have the 40-B mail plane. It could be modified to handle a couple of passengers and still have room for mail. The 40 might do it but it needs more power. We could use the new Pratt & Whitney Wasp air-cooled engine and it is two hundred pounds lighter than the Liberty. That would give you two hundred more pounds of mail."

"Good." Eddie was eager. "I think we can get our costs way under what the Post Office allows."

"We'll have to talk to Bill," said Claire.

Egtvedt decided on a plan of attack. "I'll work on the design and operating costs. You work out personnel and maintenance costs. After we put our figures together we'll have something to show Bill."

Two days later Hubbard and Egtvedt went to See Bill Boeing at his Hoge Building office in downtown Seattle and laid out their proposal.

"This is foreign to our experience," said Boeing.

"I've flown over 150,000 miles on the Victoria route without any trouble and made money," said Hubbard.

They went over the figures again. When there was nothing left to say, Egtvedt and Hubbard left.

"Oh well, Eddie it was a good try," said Egtvedt.

Bill Boeing was up for an early breakfast the next day and mentioned the meeting with Hubbard and Egtvedt, to his wife about bidding on the air mail contract from San Francisco to Chicago.

"Why not? It will develop a market for your planes," his wife replied.

Egtvedt arrived at the plant at 7:30 a.m. when the telephone operator said, "Call Mr. Boeing right away. He's been trying to get you for half an hour."

Egtvedt hurried to the phone. "Get Hubbard and come up here," said Boeing, "I want to talk some more about that proposition. It kept me awake all night."

When they got to the Hoge Building. Boeing wanted to go over the figures again, wanted to know about the bond that would be required and to compare Hubbard's costs with the Post Office figures. Hubbard planned to base the bid on a bigger load than the Post Office had been carrying. Air mail weight had been increasing and he believed publicity would increase it more. Also there was now an opportunity to carry passengers and express, which the Post Office mail planes did not carry. The plan included building their own fleet of twenty-four planes to be ready by July 1, 1927. Actually twenty-five 40A's were built with one sold to Pratt & Whitney for engine testing.

The Post Office would allow up to $3.00 a pound for the first one thousand miles and thirty cents for each additional one hundred miles. The figure Hubbard and Egtvedt came up with was $1.50 per pound for the first one thousand and fifteen cents for each additional hundred.

"Those figures look all right to me," said Boeing. "Let's send them in."

Eddie all set to test the 40-B prototype at Sand Point air field, July 1925.

Boeing officials greeting Eddie after the first test run of the 40-B mail plane.

Eddie in high spirits after first test in the Boeing Model 40-B prototype. Boeing Vice President Claire Egtvedt in straw hat shares Eddie's enthusiasm.

On January 28,1927 word was received that they were low bidders for CAM 18 which was the Post Office's designation for Contract Air Mail Route No. 18. The nearest bid was Western Air Express at $2.24 using twenty-five Douglas planes. Stout Air Service, Detroit (Ford) came in at $2.64. The Postmaster General was concerned about Hubbard and Boeing's low bid and didn't want a bankrupt carrier on his hands, particularly at the start of this most important transcontinental air mail contract. The bid was in the name of Edward Hubbard and Boeing Airplane Company. The Post Office demanded a $500,000 bond to insure performance of the contract. Bill Boeing personally underwrote the bond and a four-year contract was signed.

Boeing and Hubbard Are Low Bidders For Chicago-San Francisco Air Mail Contract

Contract Air Mail Route No. 18, or CAM 18, was between the Post Office, Boeing Airplane Company Inc., and Edward Hubbard with the United States of the first part and Boeing Air Transport, Inc. as the second part. P.G. (Phil) Johnson signed for the Boeing Airplane Company Inc., and Edward Hubbard signed as Contractor. W.E. Boeing signed for Boeing Air Transport Inc., as subcontractor.

Vice President Claire Egtvedt announced the Boeing Airplane Company would enter the commercial aviation field for the first time. To start with there would be the construction of twenty-five mail planes and then the company would probably start to manufacture similar planes for sale to the general public.

"We have long contemplated embarking in the commercial field as our previous production was mainly for the government. We lacked a starting point for such an undertaking. This contract, it seems to me, will give us our start," he stated.

The man that started all this was Eddie Hubbard. He talked Claire Egtvedt and Bill Boeing into going after the air mail contract, then he calculated the successful bid. It was also Eddie Hubbard's idea to develop the Boeing 40-B into not only a mail plane but to include a cabin for passengers. Another idea Eddie had was to build the 40-B with a metal body. *It can be said the one person that put Boeing Airplane Company into manufacturing commercial aircraft was Eddie Hubbard.*

The original Model 40 was redesigned to take advantage of the improved Pratt & Whitney "Wasp" air-cooled engine which was not only more powerful than the Liberty but was 200 pounds lighter giving more payload. This was a big plus, allowing increased revenue from two passengers plus lower operating and maintenance costs from the new engine.

Hubbard came up with the winning bid on the basis that the air mail would become more popular as would air travel. He was right on both counts. As a result they made money in the first month and never looked back. Mr. Edward Hubbard was not only one of America's best pilots; he was also a very astute business man.

Bill Boeing formed a new company to handle the air mail contract. It was Boeing Air Transport Inc., with headquarters in Salt Lake City. Eddie Hubbard was Vice President in charge of operations. The other officers were Bill Boeing, President; Phil Johnson, Vice President; Orville W. Tupper, Secretary; Claire Egtvedt, Treasurer and Willard Herron, formerly of the Seattle Chamber of Commerce as Traffic Manager in San Francisco.

Executive officers of Boeing Air Transport Inc., standing,: Claire Egtvedt, Treasurer; Eddie Hubbard, Vice President, Operations; Willard Herron, Vice President, Traffic Manager; Orville W. Tupper, Secretary; seated Phil Johnson, Vice President; William E. Boeing, President.

In March, 1927 Eddie got off a plane in Salt Lake City and received the red carpet treatment. After inspecting the airport facilities and landing field he said at a Chamber of Commerce luncheon, "This city holds the very necessary natural geographic location as an air center and will be our headquarters. There is only one other city in the United States occupying the same position in the air, and that is Chicago. "Hubbard went on to delight the city businessmen by stating his company would be carrying passengers as well as mail and estimated the cost to passengers would be about ten cents a mile and might not be a great deal higher than the railroad.

In May Eddie was giving the 40A flight trials at Sand Point air field. Marie Dunbar, a writer for the *Seattle P.I.,* went to the field to observe one of the flights. The newspaper language of the day was, to say the least, flowery. This is one of her observations, "The silvery, shiny giant of the skies quivered like a fluttering bird and then rose with stately sureness." Some of her other observations about this shiny giant: "This machine has a self starter. The passenger cabin is heated. There is a fire extinguisher, there are shock absorbers, brakes to control the wheels and lighting equipment to make night flying possible." I would imagine Eddie outlined all the new features for Miss Dunbar.

Seattle Post-Intelligencer

SEATTLE, JUNE 17, 1927.

Here Is A Fortune On Wings

FIVE TIMES $22,000 = $110,000

Five of the sixteen Boeing air mail planes lined up ready for flight tests at Sand Point Field. Each ship, designed to carry two passengers in addition to the pilot and mail, was fabricated in the Boeing plant down to the last strut, with the exception of engine, and represents an investment of $22,000. Twenty-five of such planes, all products of the Boeing factory, will be on the Chicago-San Francisco air mail run by July 1, President P. G. Johnson of the company, announced yesterday.—(Post-Intelligencer Photo.)

Boeing 40-B mail and passenger planes lined up at Sand Point air field before dispersed along the air mail route between San Francisco and Chicago.

By the middle of June 1927 three of the 40A mail planes had flown to Salt Lake City in preparation of the start of service on the 2,000 mile run between San Francisco and Chicago, July 1st. They made the 800 mile trip in 7 hours. The other mail planes were flown to San Francisco and other points. Eddie already had flown one to Chicago.

Standard Furniture Co., Seattle, had a large ad in the newspapers with a picture of a Boeing 40A, "For The First Time - The New Boeing Mail Plane On Public Exhibition At Standard Furniture (Main Floor)." Several days later it was reported ten thousand people had visited the store in two days.

THE SEATTLE DAILY TIMES — SUNDAY MORNING, JUNE 19, 1927.

On Exhibition for a Few Days Only! (MAIN FLOOR)

The Boeing Mail Plane — Achievement of Seattle's Largest Manufacturing Plant!

For the First Time—
The New Boeing Mail Plane on Public Exhibition at Standard Furniture Co. (Main Floor)

Features of Interest About This Mail Plane!

Over 750 expert mechanics, engineers, draftsmen, etc., are employed by the Boeing Air Plane Co. in its construction and that of other air planes now being produced by this large Seattle Plant.

The Boeing Mail Plane will carry 1,[illegible] lbs. of mail and two passengers or 1,[illegible] lbs. of mail.

It has a flying radius of [illegible] miles, carries 100 gallons of gasoline—60 gallons in the body and [illegible] gallons in the upper wing.

The engine is equipped with an electric starter. An engine driven generator and storage battery provide electric current for starting and lighting.

To prevent fire there is a pressure fire extinguishing system in engine compartment, controlled from pilot's seat. Weight with gasoline, passengers and mail is 5,200 lbs.

The span of the wings is 44 ft. 2½ in. and length of body over all 33 ft. 3 in.

The engine is a Pratt and Whitney "Wasp" of 410 horsepower. The propeller is made of "Standard Steel," forged duralumin.

It has a maximum speed of 135 miles an hour. And can climb to an altitude of 15,000 feet—the first 10,000 feet in 12.5 minutes.

It is one of 25 similar Mail Planes being operated by the Boeing Air Transport Co. between San Francisco and Chicago — the Main Stem of the air mail transport system —representing an investment of close on to $1,000,000 of Seattle capital.

This entire Mail Plane is now on display on the Main Floor of the STANDARD.

THROUGH the courtesy of the Boeing Air Plane Co. of Seattle the STANDARD takes pride and pleasure in announcing an exhibition, in its entirety, of one of the identical 25 new Boeing Mail Planes to be operated by the Boeing Air Transport, Inc., on the division between San Francisco and Chicago, the Main Stem of the United States Transcontinental Air Mail Route.

THIS very latest creation in Airplane Engineering—the product of Seattle's largest manufacturing plant—in an exhibition requiring very near our entire Main Floor—is a feature all are cordially invited to visit, June 20-25 inclusive.

Transcontinental Air Mail Route

This Map shows the route of the United States Transcontinental Air Mail over which you can now send a half ounce of mail to any point in the United States for 10c. Your letter from Seattle will be delivered to New York in 45 hours—to Chicago in 35 hours. (Planes are in operation night and day.

STANDARD FURNITURE CO.
L. SCHOENFELD & SONS
SECOND AVENUE AT PINE STREET

Three days before taking over the San Francisco-Chicago air mail route from the Post Office, Bill Boeing, Phil Johnson, and Eddie Hubbard took off from Sand Point in one of the new 40A's to survey part of the route. With Eddie as pilot they flew to Salt Lake. Boeing went to San Francisco to see the first flight for the East. Mrs. Boeing met him and christened the first plane for a venture she convinced her husband to take. Hubbard and Johnson flew to Chicago, stopping at their terminal points along the way.

With the service set to go on July 1, 1927 Eddie made the following announcement: "The new Boeing air mail planes are equipped to carry two passengers and mail including baggage up to 1,600 pounds. The time from San Francisco to Chicago will be 22 1/2 hours. Twenty-five planes will be used doing 130 miles an hour. The aircraft have a new 420 horsepower Pratt & Whitney 'Wasp' air cooled engine. Each mail plane was built at a cost of $20,000, making Boeing Air Transport's investment $500,000. The San Francisco-Chicago line will make passenger connections with the Los Angeles-Seattle airline at San Francisco. With the Salt Lake City-Pasco line at Salt Lake and the Cheyenne-Pueblo line at Cheyenne. We are working on reciprocal arrangements between various airlines where passengers can buy through tickets."

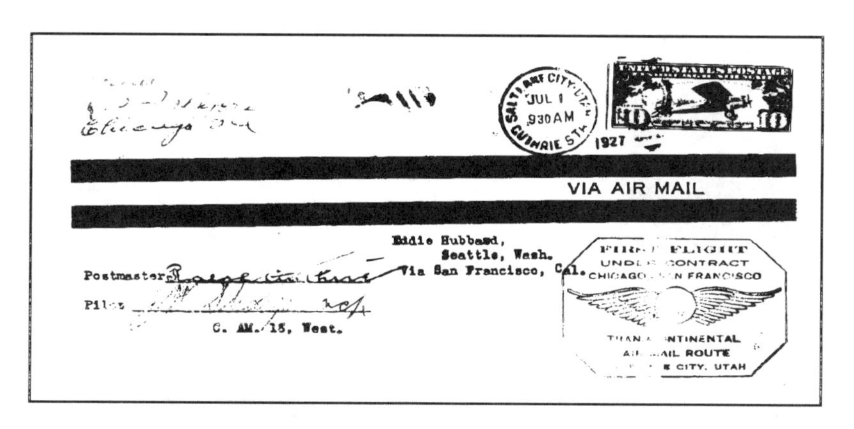

FIRST FLIGHT COVER SALT LAKE CITY TO SAN FRANCISCO. CONTRACT AIR ROUTE 18, JULY 1, 1927.

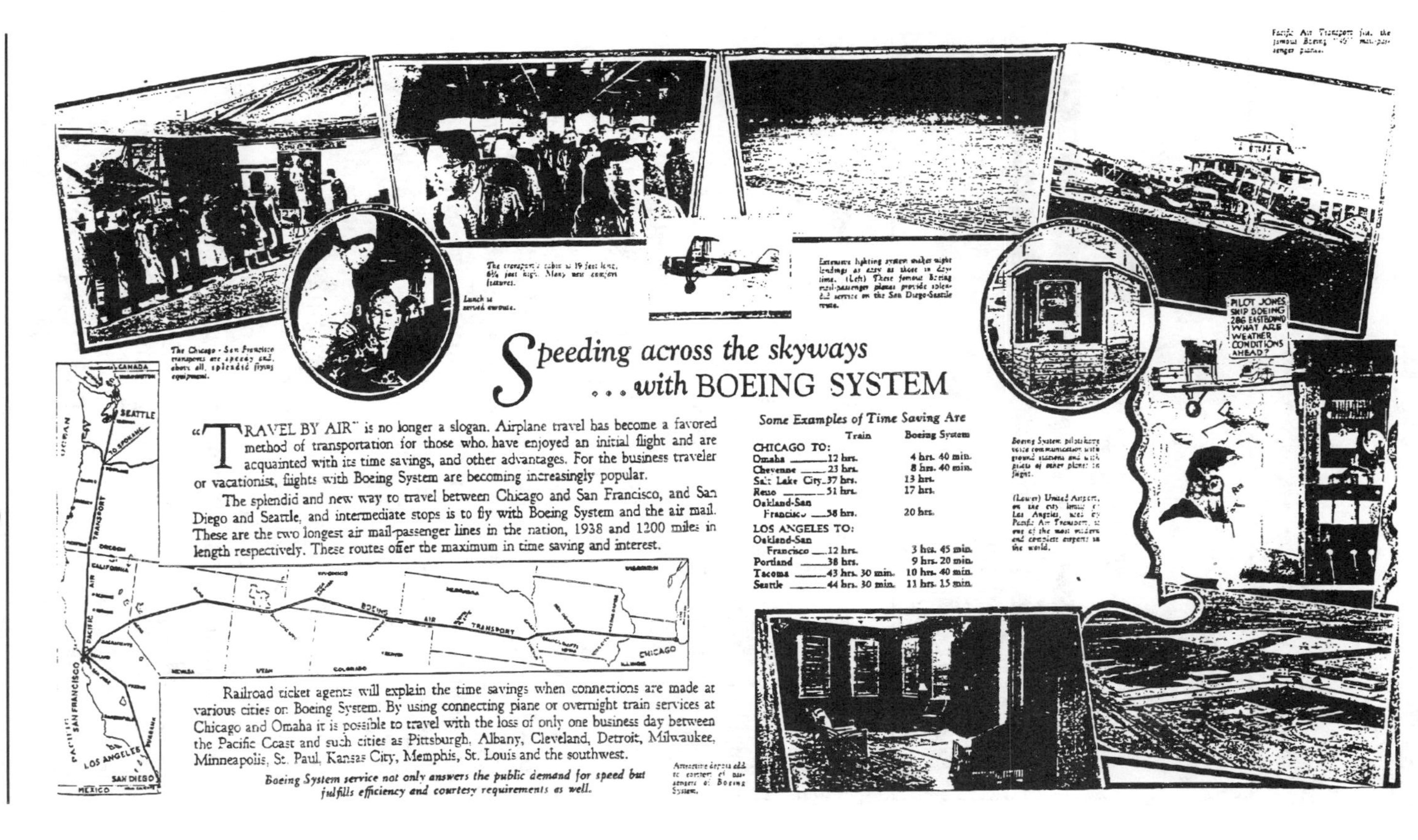

The Chicago-San Francisco transports are speedy and, above all, splendid flying equipment.

Lunch is served enroute.

The transport's cabin is 19 feet long, 6½ feet high. Many new comfort features.

Extensive lighting system makes night landings as easy as those in day time. (Left) These famous Boeing mail-passenger planes provide splendid service on the San Diego-Seattle route.

Pacific Air Transport [illegible] the famous Boeing "[illegible]" mail-passenger planes.

Speeding across the skyways ... with BOEING SYSTEM

"TRAVEL BY AIR" is no longer a slogan. Airplane travel has become a favored method of transportation for those who have enjoyed an initial flight and are acquainted with its time savings, and other advantages. For the business traveler or vacationist, flights with Boeing System are becoming increasingly popular.

The splendid and new way to travel between Chicago and San Francisco, and San Diego and Seattle, and intermediate stops is to fly with Boeing System and the air mail. These are the two longest air mail-passenger lines in the nation, 1938 and 1200 miles in length respectively. These routes offer the maximum in time saving and interest.

Some Examples of Time Saving Are

	Train	Boeing System
CHICAGO TO:		
Omaha	12 hrs.	4 hrs. 40 min.
Cheyenne	23 hrs.	8 hrs. 40 min.
Salt Lake City	37 hrs.	13 hrs.
Reno	51 hrs.	17 hrs.
Oakland-San Francisco	58 hrs.	20 hrs.
LOS ANGELES TO:		
Oakland-San Francisco	12 hrs.	3 hrs. 45 min.
Portland	38 hrs.	9 hrs. 20 min.
Tacoma	43 hrs. 30 min.	10 hrs. 40 min.
Seattle	44 hrs. 30 min.	11 hrs. 15 min.

Boeing System pilots have voice communication with ground stations and with pilots of other planes in flight.

(Lower) United Airport, on the city limits of Los Angeles, used by Pacific Air Transport, is one of the most modern and complete airports in the world.

Railroad ticket agents will explain the time savings when connections are made at various cities on Boeing System. By using connecting plane or overnight train services at Chicago and Omaha it is possible to travel with the loss of only one business day between the Pacific Coast and such cities as Pittsburgh, Albany, Cleveland, Detroit, Milwaukee, Minneapolis, St. Paul, Kansas City, Memphis, St. Louis and the southwest.

Boeing System service not only answers the public demand for speed but fulfills efficiency and courtesy requirements as well.

Attractive depots add to comfort of passengers of Boeing System.

United States Contract Air Mail Route 18 between San Francisco and Chicago.

Boeing Air Transport's first passenger was Jane Eads, a reporter for *The Chicago Herald* and *Examiner*, who flew from Chicago to San Francisco for a fare of $200. Her first stop with pilot Ira Biffle was at Iowa City for fuel and she filed her first report. Biffle was not only one of the line's best pilots, he was also Charles Lindbergh's first flying instructor. Jane filed her next report at Omaha on an empty stomach, as food facilities for passengers were yet to be organized. After a night's flight one of her comments was, "Towns are nothing but lights!" Miss Eads was greeted by Eddie Hubbard and E.B. Wadsworth, superintendent in charge of contract air mail.

Soon she was airborne again on her way west with one of America's best known air mail pilots, Jack Knight. Over 300 pounds of the first mail carried west from Chicago was official and personal letters consigned to President and Mrs. Coolidge. These mail bags were transferred to an army plane at North Platte a fueling stop in Nebraska.

The trip to San Francisco took a little over 22 hours, which was 40 hours faster than train. Phil Johnson, Boeing President, reported twelve days after taking over the route, "Everything is going smoothly and we haven't had a single delay. The volume of mail is increasing and many applications are being received for passenger tickets. We surprised the world by taking over the route on the day we were supposed to." *This was a great compliment to Eddie Hubbard. The man who made it all happen.*

During the first month of operation, July 1927, George Mead, Pratt & Whitney Vice President, decided to review his troops stationed at Salt Lake City. These men looked after the 'Wasp' air-cooled engines used by Boeing Air Transport in their 40-B mail and passenger planes.

Some of the landing fields were not as smooth as silk and he was amazed how the new Boeing Oleo landing gear handled the mounds of sand created by sagebrush. This gear made landing and taking off much smoother.

George came west and met Eddie Hubbard at Salt Lake City. Hubbard had just flown in from San Francisco to keep a dinner engagement with Mead, who was very impressed, but a six-hundred-mile flight to Eddie Hubbard was all in a day's work. Mead spent two days in Salt Lake City observing the Boeing Air Transport operation and was impressed. He was able to meet Hubbard's General Superintendent, D.B. Colyer, ex-Army flyer and Second Assistant Postmaster General and Eastern Division Superintendent, Caldwell. Colyer was a very valuable asset to Hubbard, as he knew all the Post Office pilots and hired the veterans who had been flying the route by day and by night for the past three years.

George made an observation about W.E. Boeing, echoed by many, "It is no wonder that he sees visions and makes them become realities."

After Salt Lake City the trip west to San Francisco was in five hundred mile hops with a fresh pilot for each leg of the journey. Salt Lake was a busy divisional point. Western Air Express, who had an air mail contract from Salt Lake City to Los Angeles and Varney Air Lines, who had an air mail contract from Pasco, Washington to Elko, Nevada with a stop at Boise, Idaho. Both had hangars at the Salt Lake City airport.

It took four different pilots to fly the 2,000 miles from Chicago to San Francisco. The first leg was Chicago to Omaha with a stop at Des Moines. The second leg was to Cheyenne with a stop at North Platte. The third leg was to Salt Lake City with a stop at Rock Springs. The fourth to San Francisco with stops at Elko, Reno and Sacramento.

Figures released for August, the second month of operation, were:

- Miles flown 121,949
- Hours flown 1,217
- Passengers carried 122

The air mail poundage was steadily increasing. Boeing Air Transport was proving Eddie Hubbard's theory that cross-country air mail and passenger service was a profitable venture without subsidy.

American Railway Express making delivery in a Model T Ford truck to Boeing Air Transport 40-B mail and passenger plane.

DES MOINES SUNDAY REGISTER—OCT. 28, 19

WEEKLY PAGE OF AVIATION NEWS

Loading the Mail at the Chicago Airport

Speed is the keynote of air transport service. This photo, taken at the Chicago airport, shows how the work is systematized. The planes, dropping in from all parts of the nation with mail and express, taxi up to the platform in the foreground to discharge their cargo and take on new loads. The trucks, after turning the outgoing mail over to the planes, rush the Chicago mail back to the postoffice for distribution.

In September 1927 officials of American Railway Express Company, later known as American Express, came to Seattle and made arrangements with Boeing Air Transport to handle air express on the Chicago - San Francisco route. The first day was so successful a second plane had to be put into service. The first plane carried seventy-six express packages. One American Express official predicted larger planes in the future. He must have been reading Phil Johnson's mind, as the much larger Boeing trimotor Model 80 was about to enter the scene.

1927 had been a great year for aviation. In May Charles Lindbergh had flown the Atlantic non-stop. In August the Dole Race for $25,000 was completed from Oakland, California to Wheeler Field, Oahu, Hawaii.

In January 1928 Boeing Air Transport purchased a substantial interest in Pacific Air Transport, Vern Gorst's company, which operated the air mail and passenger route between Seattle and Los Angeles. Seattle was now the terminus of the longest air passenger service under one control in the world.

For sometime, Phil Johnson, Boeing President had been developing a larger aircraft for their San Francisco-Chicago route. With the build-up of the number of passengers he could see that not too long down the road, carrying passengers would be the top priority and mail would take second place. From the Boeing factory came a twelve passenger tri-motor aircraft with comforts the flying public had never experienced. Some airlines in Europe were using male stewards. Now Boeing went a step further with the world's first stewardesses: they all had to be nurses. The aircraft was very advanced for its day. Reclining and upholstered seats, heated cabin, reading lamps, box lunches, lavatory, hot and cold running water. The cabin was insulated and the passengers were no longer deafened by the engines roar. As a result of these new innovations the aircraft became known as "the pullman of the air."

Boeing Air Transport, Inc. new Model 80 for use on United States Contract Air Mail Route 18 between San Francisco and Chicago.

Another of Eddie Hubbard's predictions had arrived. When he and Bill Boeing had returned to Seattle from the 1920 aircraft manufacturers' show in San Francisco he told the press that eventually great strides would be made in passenger comfort.

It was now time to present the Model 80. When this larger aircraft was unveiled they realized that all the airfields they used had to have long enough runways to handle the bigger plane. One field was too short and that was Reno's. Eddie was sent to Reno to see what could be done. He found that the present landing field was too short and close to the mountains. It was also in a very windy spot. A few weeks before he arrived, Charles Lindberg landed in Reno and made the same observations. After his report, Bill Boeing told Eddie to find a new site, buy it and start building a hangar.

Eddie purchased 120 acres in the center of the valley and made arrangements for a hanger to be constructed. Hubbard announced to the happy citizens that neither the city nor the county would be asked to participate in the purchase of the field or in the construction of the hangar. Boeing would manage the field and it would be open to all planes. The site would be ready to handle the new 12 passenger planes within sixty days.

Boeing 40-B mail plane at Hubbard Field, Reno, Nevada, in front of hangar built by Boeing Air Transport Inc. This is the site of the present Reno Cannon International Airport.

What a gift to Reno. It also made the old site available for the city to sell. The original airfield was Blanche Field, named after a local air mail pilot. They planned to move the existing hangar to the new location. Hubbard also added, "The new field is level and dry. With a little work it will be as good as any in the country and will be able to handle any size of plane with safety. The Boeing Company expects to invest about $50,000 in the development. The land cost $150 an acre and the hangar will cost over $25,000."

Boeing Air Transport announced that the largest volume passenger traffic was between Oakland and Reno. Of the 181 passengers on this short haul only 27 went all the way to Chicago. The report also said, "Perhaps urged by traffic congestion on the ground, residents of big cities take to the air." I wonder what they would think of today's traffic congestion!

On July 25,1928 the United States Post Office cut air mail postage from ten to five cents per ounce. By August 12th air mail arriving and leaving Seattle had more than doubled from 20,000 envelopes and postcards to 40,000 a day. This is exactly what Eddie Hubbard predicted the previous year when he did the calculations for Boeing Air Transport's successful bid. Air mail poundage kept on increasing, giving a greater return on Boeing's investment.

Boeing President Phil Johnson believed air traffic would rapidly increase, requiring larger aircraft. He was the man who pushed for a jump from the two-passenger 40-B to the trimotor Model 80 that carried twelve passengers and mail. A later version the 80-B carried eighteen passengers and mail.

Passengers waiting to board United Air Lines Boeing Trimotor Model 80. This aircraft had comforts air travellers were not used to -- the first stewardesses, heated cabin, upholstered and reclining seats, forced air ventilation, hot and cold running water, lavatory, insulated and soundproof cabin, reading lamps and box lunches.

In late November the Model 80 arrived at the new field in Reno with eight passengers, 785 pounds of mail and a large quantity of express and baggage. Most of the passengers on this first Model 80 flight were Boeing Air Transport officials. At the airport opening ceremonies Bill Boeing announced to one and all, "I now christen this new landing site 'Hubbard Field?"

Twenty pounds of mail was added to the largest that had been carried over the Sierra making a total of 805 pounds when the plane left Hubbard Field. The 120 acres originally named Hubbard Field was acquired from Boeing Air Transport Inc., in 1936 by United Air Lines who continued to own and operate the airport until 1953 when the City of Reno purchased it. In 1979 the airport's name was changed to Reno Cannon International Airport after Senator Howard W. Cannon of Nevada.

Boeing Air Transport received the following letter from a very unhappy Reno air mail customer.

Dear Sir:
I just want to know what sort of a rotten airplane you got. I sent a letter two weeks ago to my daughter who was going to be married to a man in St. Paul. I went to the trouble and expense of five cents to send this letter by air mail. They told me the letter would get there the day after, but it ain't there yet. I could make better time with a horse. In fact, I have often done so. Now I ain't going to holler, but I want you to find out what's holding this letter back. My daughter thinks I'm a liar.
P.S. Since I wrote this I have found my daughter's letter in a pocket of my blue pants.

Now there was one honest man!

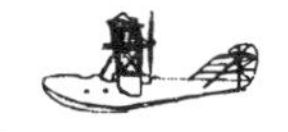

Chapter 20
Tributes to a Great Aviator and Airline Builder

VICTORIA DAILY TIMES

Famous Aviator Succumbs After Brilliant Record

Known From Victoria to the Golden Gate as the Flying Postman, "Eddie" Hubbard, Pioneer of Seattle-Victoria Air Mail Flights, Dies at Salt Lake City After Operation; Had Eighteen Years Unbroken Record for Safe Air Navigation

ACTIVITIES END

Edward Hubbard, Boeing Air Transport officer, who died Tuesday.

HUBBARD, AIR LINE BUILDER, PASSES AWAY

Death Comes in Salt Lake Hospital, Following an Operation.

DEATH FOLLOWS OPERATION ON 'EDDY' HUBBARD

Aviator Who Inaugurated First Air Service to This City, Dies in Hospital in Salt Lake City

HELD SPLENDID RECORD AS FLYER

DEATH PILOTS ED HUBBARD ON LAST HOP

FIRST MAIL FLYER

VETERAN FLYER HUBBARD DIES IN SALT LAKE CITY

Uncle Sam's First International Air Mail Pilot and Boeing Transport Vice President Succumbs.

The world of aviation was stunned on December 18, 1928 by the death of Eddie Hubbard at the early age of 39. Eddie Hubbard went into Salt Lake City's Holy Cross hospital for a stomach ulcer operation Sunday, December 16, 1928 and died Tuesday, December 18th, of an infection. Although his condition was known to be fatal he talked to Boeing associates and arranged his will with attorneys before passing away that evening.

None of his friends had been advised of his illness. Not too long before, his appendix was removed at Omaha and he appeared to be in good health. For a man who so many times faced death this was an ironic twist of fate.

To attest to Hubbard's popularity the *Seattle Post Intelligencer* carried the following headline, "DEATH PILOTS ED HUBBARD ON LAST HOP." The *Victoria Daily Times* had his picture and a write-up on the front page with a headline, "FAMOUS AVIATOR SUCCUMBS AFTER BRILLIANT RECORD." *The Victoria Colonist*: "DEATH FOLLOWS OPERATION ON EDDIE HUBBARD." *The Seattle Times*: "VETERAN FLYER HUBBARD DIES IN SALT LAKE CITY." As further tributes to this great pilot and airline builder the *Seattle Post Intelligencer*, *Victoria Daily Times* and *Salt Lake Tribune* had the following editorials.

EDWARD HUBBARD SOARS ON

Seattle Post Intelligencer
Thursday, December 20, 1928

The whole world of aviation, celebrating the twenty-fifth anniversary of the science, will sorrow at the death of Edward Hubbard.

"Eddie's" fifteen years of active flying compassed almost the whole great epic of aviation. He took to the air in 1913, when boys did not so avidly put on wings, and the air remained the element of his being until he died.

He established and flew the first contract air mail route in the world. From this start — made between this city and Victoria, B.C. — carrying of mail by air has spread to the entire civilized world.

By his ability to hold regularly to schedules, winter and summer, on his Seattle-Victoria flight, "Eddie" Hubbard enlisted early interest by the federal government and thereby was personally responsible for a vast practical contribution to flying.

He was one of the few pilots who at the entry of this country into World War was competent to instruct army fliers. In his knowledge and experience his was the strength of regiments.

His long career in the air is eloquent of his mastery of technique. His abilities were a continuous source of inspiration to William E. Boeing and played some part in influencing that leader to develop in Seattle the greatest of America's aircraft building plants. In this concern Hubbard rose to the position of operating official.

Hubbard died at thirty-nine from effects of an operation. His going, was not the swift, heroic end which we associate with aviation. But his whole mature life was filled with practical heroisms which have been so influential in developing commercial aviation.

EDDIE HUBBARD

Editorial - *Salt Lake Tribune*, December 18, 1928.

Eddie Hubbard is dead in Salt Lake City and there will be mourning among the children of Victoria and Seattle when they hear the news. They still remember him, a great many of those children, and how he used to fly in from the sea, and sweep over the city in his great plane, carrying the mails to and from the Orient liners when they put in at the Outer Wharf at Victoria, down by Dallas Road. He was a great airman, was Eddie Hubbard, and he was a hero to the children, and his name was a legend among them. Even today, when the mail plane from Seattle comes in, although he has not been flying it himself for the last two years or so, the Victoria school children will tell you: "That's Eddie's plane."

Eddie Hubbard will come down to the sea from the air no more, and he died in his bed, but he wrote a splendid chapter of his own in the adventure of the airways. He was the first aerial postman of the United States, and he made his first contract to carry the Pacific mails between Seattle and Victoria more than eight years ago, and he flew his seaplane himself in that service for six years. He established a great record. He never missed a ship or lost a letter. Rain or shine, storm or mist, he came across the waters of the Strait to meet the liners. Once he made five trips between the two ports in twenty-four hours. He had many an adventure, and the luck of the brave must have been with him in many a last chance; but the legend of his charmed airmanship came to be invincible against the threat of misadventure, and it almost seemed in Seattle and Victoria, at least, as if Eddie Hubbard must be the airman immortal.

The news stories about him tell us that his record as a mail carrier of the air stands without parallel, and that "probably nowhere in the world has an air pilot in the mail service ever done such extracting service." That may well be true, although we do not know whether it is precisely so or not. It does not matter. It is enough that he survived the chances of the airways for fifteen years, and that his achievement was a real contribution to the art and science of aerial navigation. And what we like to hear best of all about him in his obituary notice is that "Eddie Hubbard worked without faltering." It is still meant that we should celebrate the pioneers of the air, and it becomes us that we should pay our tribute to one of the best of them, Eddie Hubbard, who flew the air mail to Victoria where the children watched to see him come in from the sea.

THE LATE "EDDIE" HUBBARD

Victoria Daily Times, December 20, 1928.

Just a little more than eight years ago a young aviator conceived the idea that he might persuade the United States Government to allow him to inaugurate the first aerial mail service on this continent. He explained to the authorities at Washington that by linking Victoria and the Sound City by air, businessmen of this coast would be able to get quicker delivery of mail from the Orient and wait until almost sailing time before closing their outbound mail. The suggestion appealed to them and a contract was entered into. The man who flew with the first bag of mail in his official capacity as a government contractor was Edward Hubbard. But "Eddie" — as he was affectionately called by a host of friends in this city and in other parts of the continent — will never fly again. He died at a hospital in Salt Lake City on Tuesday night after an operation for ulcer of the stomach. He was thirty-nine years old. "Eddie" Hubbard did not fly across the Atlantic. He attempted neither altitude nor endurance records. Stunting was not included in his flying performance. He was one of the pioneers of commercial aviation on this continent, the first postman of the air, first in the international service in the New World — and he never lost a bag of mail or met with anything more serious than one or two forced landings. For six of the eight years during which the Seattle-Victoria contract has been in effect "Eddie" flew the mail plane himself and his comings and goings, although they soon lost their novelty, always interested the people of this city. They were signals to those not generally familiar with the movements of transpacific shipping that a liner either was about to arrive or depart. "Eddie" was on the job.

> As we have said, however, he was not a spectacular airman; his name never became a household word and there is no chance of that now. But if posterity is fair to "Eddie" Hubbard, it will say of him that he was a really great airman and a successful pioneer in the realm of practical aviation.

To indicate Eddie Hubbard's admiration and popularity in Victoria the above was the second bereavement on the *Times* editorial page that day. Eddie shared his obituary with that of the Hon. Walter C. Nicol, Lieutenant Governor of the Province of British Columbia.

Twenty-one years after his passing the *Victoria Daily Colonist* had an editorial about Seattle-Victoria Air Links... "Victorians have not forgotten the pioneering of Mr. W.E. Boeing and his chief pilot, the late "Eddie Hubbard". None the less the Hubbard contract, seven years in extent, proved the worth of civil flight in this region and before it was established anywhere else on the continent ... At this moment, both Seattle and Victoria will be thinking of a wiry man, with a wide grin and steady eyes — the late Eddie Hubbard, our first postman by air. He would approve."

His long time friend and employer, Bill Boeing, had this to say, "Eddie Hubbard was a clear thinker, kind to everyone beloved of all who knew him, fair and just and an inspiration to all who came into contact with him. His death was unexpected. Our loss is very great and we feel very keenly the severing of longtime association. Under his direction the transport service was most efficiently rounded out and brought to the high standard under which it is now operated."

A resolution was passed by the Salt Lake City Chamber of Commerce expressing its appreciation and respect for the services rendered by Edward Hubbard, his high ideals, his status as good citizen and the courageous and efficient administration of his duties. Also a letter of condolence was sent to Mrs. Hubbard.

The American Society of Mechanical Engineers

UTAH SECTION

Executive Committee

Dec. 22, 1928.

Mrs. Mildred Hubbard:
℅ Boeing Airplane Co.,
Seattle, Wash.

My dear Mrs. Hubbard:

The members of the Utah Section of the American Society of Mechanical Engineers wish to express their deep sympathy to you for the loss of your husband and our friend, Eddie Hubbard.

We had come to know and to like him after he appeared before our Society to tell us of the Recent Developments in the Air Craft Industry.

It is a matter of deep regret to us all that such a promising career as Eddie's could not have been permitted to unfold its great possibilities and to achieve its great reward.

Very sincerely yours,
Utah Section A.S.M.E.
by Walter H. Trask Jr.
Chairman

WRIGHT AERONAUTICAL CORPORATION
PATERSON, N. J.
U. S. A.

January 15th, 1929.

My dear Mrs. Hubbard:--

I learned only last night, from Mr. Herron, of Eddie's death. Even this morning I can hardly believe it is true.

May I offer you my most sincero sympathy in this loss, as well as to express to you the similar feelings held by everyone in this Company who had the privilege of knowing your husband.

To me, Eddie's attributes placed him in a unique position in Aeronautics. I have known many good pilots and operation managers, but none who were comparable at all with him. His balance, his judgement, and his personal charm placed him alone, in my estimation.

The facetious remark by which Phil Johnson spoke of Eddie, "he is the only pilot who can fly, keep a level head and keep his feet on the ground all at the same time", seemed to typify your husband. He was the only man to whom I have ever felt that such an expression could be applied.

Again let me tell you how extremely sorry I am to learn of this loss. It is not only your loss. It is a loss to all Aeronautics.

Very sincerely yours,
J. T. Hartson.

JTH:

Mrs. Edward Hubbard,
1215 Franklin Ave.,
Salt Lake City, Utah.

USE THE AIR MAIL

Early in December Eddie prepared a paper for the Society of Automotive Engineers. The Chairman was Glenn L. Martin of aircraft manufacturing fame. Evidently the Society was not aware of Eddie's passing and sent him the following letter of appreciation.

Eddie Hubbard, who had been with Boeing in the lean days, must have been pleasantly surprised a month before he passed away when his close friend of many years had gone public with a share issue putting an infusion of $4,000,000 cash into the company. Net earnings for 1928 were estimated at $1,300,000 and the company was now employing 1,000.

SOCIETY OF AUTOMOTIVE ENGINEERS
INC.
TWENTY NINE WEST THIRTY NINTH STREET
NEW YORK

December
20
1928

Mr. Edward Hubbard,
Boeing Airplane Co.,
Seattle, Washington.

Dear Mr. Hubbard:

I wish to express my appreciation of the excellent paper, prepared by you and Mr. L. S. Hobbs, for our Aeronautic Meeting that was held in Chicago early this month. The paper, which was presented at the Power Plant Session, was a distinct contribution to the success of the meeting, and we are sorry that you were not there to see how well it was received. We hope to have you with us at future Aeronautic meetings.

I am very appreciative of your courteous cooperation and wish to thank you on behalf of Chairman Glenn L. Martin and the members of the Aeronautic Meeting Committee, as well as on my own behalf.

Yours very truly,

SOCIETY OF AUTOMOTIVE ENGINEERS, Inc.

Manager, Meetings Department

Aeronautic Meeting
Technical Program Committee:
Glenn L. Martin, Chairman
Hon. E. P. Warner
Capt. E. S. Land
Capt. L. M. Woolson
Earl D. Osborn

C.Heywood/EB.

WHEN YOU NEED AUTOMOTIVE ENGINEERS USE THE S.A.E. EMPLOYMENT SERVICE

The above expressions are but a few of the outstanding qualities his peers saw in him. Eddie is buried at Evergreen-Washelli cemetery on Aurora Avenue, Seattle.

Much to the surprise of many people, when Eddie's will was probated at Superior Court in Seattle, he left his widow Mildred and his stepdaughter Margaret $1,200,000. Margaret later became Mrs. John Graham residing on Bainbridge Island. It was mentioned before that Eddie was not only an outstanding pilot but was also an astute businessman. He invested in real estate, stock in the Boeing Airplane Company, Pacific National Bank and Boeing Airplane & Transport Corporation. Pacific National Bank was the creation of Bill Boeing.

MILLIONS AND DEATH COME SAME TIME

Notable Career of Seattle's Beloved Aviator Had Lowly Beginning Here

$1,200,000 IN HUBBARD WILL

AIR PILOT'S WEALTH IS REVEALED IN TESTAMENT

Eddie Hubbard's grave at Evergreen-Washelli Seattle. The bowl-shaped gravestone has a reproduction of Boeing Air Transport emblem.

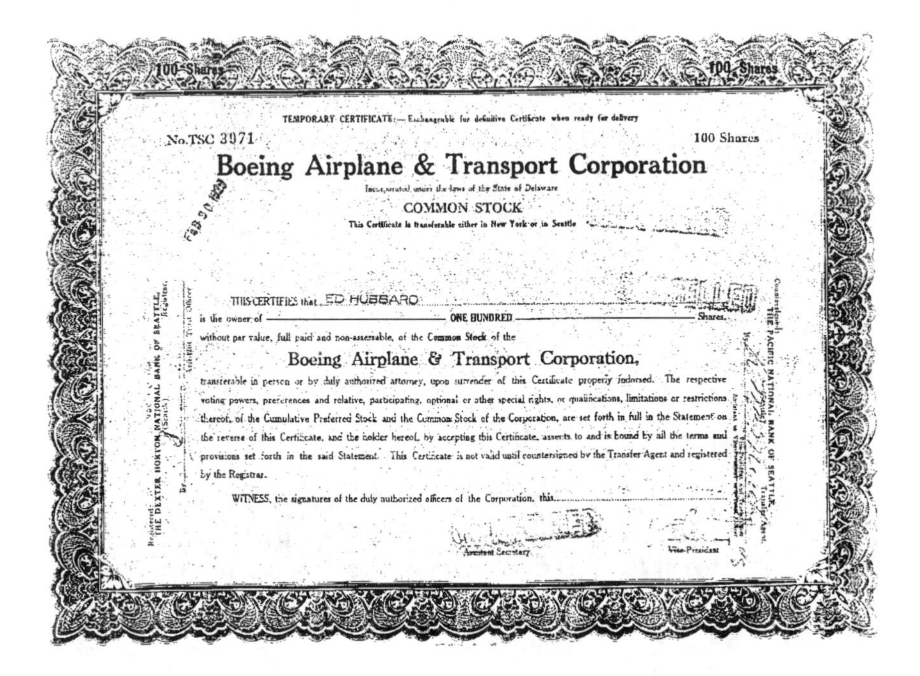

100 Shares   100 Shares

TEMPORARY CERTIFICATE:—Exchangeable for definitive Certificate when ready for delivery

No.TSC 3971   100 Shares

Boeing Airplane & Transport Corporation

Incorporated under the laws of the State of Delaware

COMMON STOCK

This Certificate is transferable either in New York or in Seattle

THIS CERTIFIES that ED HUBBARD

is the owner of ONE HUNDRED Shares,

without par value, full paid and non-assessable, of the Common Stock of the

Boeing Airplane & Transport Corporation,

transferable in person or by duly authorized attorney, upon surrender of this Certificate properly indorsed. The respective voting powers, preferences and relative, participating, optional or other special rights, or qualifications, limitations or restrictions thereof, of the Cumulative Preferred Stock and the Common Stock of the Corporation, are set forth in full in the Statement on the reverse of this Certificate, and the holder hereof, by accepting this Certificate, assents to and is bound by all the terms and provisions set forth in the said Statement. This Certificate is not valid until countersigned by the Transfer Agent and registered by the Registrar.

WITNESS, the signatures of the duly authorized officers of the Corporation, this

Assistant Secretary   Vice-President

Registered: THE DEXTER HORTON NATIONAL BANK OF SEATTLE, Registrar.

Countersigned: THE PACIFIC NATIONAL BANK OF SEATTLE, Transfer Agent.

One of Eddie's 100 share certificates in Boeing Airplane and Transport Corporation.

The career of one of America's outstanding aviators, instructors, air mail pilots and air line builders came to an abrupt close. He was much like his friend and employer, William E. Boeing, *a man of vision.* There are few famous men in aviation that have had so many editorials lauding their achievements.

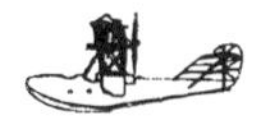

Chapter 21
Hubbard Was Predecessor of United Air Lines

A few weeks before Eddie's untimely death his mentor, Bill Boeing, formed at that time the largest conglomerate in American history — the United Aircraft and Transportation Corporation.

Again we see an example of Bill Boeing's foresight. He realized that to be successful and profitable you must grow. Being bigger had decided advantages. In 1928 he formed and became chairman of United Aircraft & Transport Corporation with Fred Rentschler of Pratt & Whitney as President; Chance Vought, head of the Chance Vought aircraft company as Vice President; Charles W. Deeds, National City Bank, Secretary-Treasurer. The impressive board of directors included Philip Johnson, President of Boeing Airplane Company and Boeing Air Transport Inc.; Charles F. Kettering, Vice President General Motors; Kenneth R. Kingsbury, President Standard Oil Company of California; Charles K. Knickerbocker, Vice President Griffin Wheel Co.; William B. Mayo, Chief Engineer, Ford Motor Company; George J. Mead, Vice President, Pratt & Whitney; Gurney Newun, Los Angeles attorney; Gordon Rentsckler, National City Bank; Joseph Ripley, National City Company and Orville Tupper, Secretary-Treasurer, Boeing Airplane Company. Under the umbrella of United Aircraft control of the following companies began:

- The Hamilton Standard Propeller Corporation.
- Hamilton Metal Plane Division.
- Chance Vought Corporation.
- The Pratt & Whitney Aircraft Company, which included Canadian Pratt & Whitney Aircraft Company Ltd.

- Sikorsky Aviation Corporation.
- Boeing Aircraft of Canada Ltd.
- The Stearman Aircraft Company.
- Pacific Air Transport, Inc. (acquired 1928)
- Stout Air Services, Inc.
- Varney Air Lines, Inc. (acquired 1930)
- National Air Transport, Inc. (acquired 1930)

Headlines across the nation proclaimed, "$150,000,000 MERGER IN AVIATION INDUSTRY." It all started when Eddie Hubbard convinced his employer and friend to go after the San Francisco-Chicago air mail contract and build their own mail planes.

By July 15, 1933 United Aircraft & Transport Corporation capitalization outstanding was preferred stock at $50 par $7,500,000 with 2,086,839 shares of common stock at no par value. To complete the picture of this gigantic jump in size in 1934 United Airports of Connecticut was acquired and later became a department of Pratt & Whitney, along with United Air Lines, Inc.

United Air Lines took control of the four airlines of United Aircraft & Transport, which were Boeing Air Transport, National Air Transport, Pacific Air Transport and Varney Air Lines.

To quote from *Corporate and Legal History of United Air Lines and its Predecessors and Subsidiaries,* edited by Adrian Delfino, in 1953: "Mr. Edward Hubbard, in a sense, was the first air transport predecessor of United Air Lines because of his contract to deliver mail from Seattle to Victoria and back and his interest in Boeing Air Transport." It was basically Boeing Air Transport that was the key link in the future United Air Lines system. United's headquarters were started in Chicago with Phil Johnson as President.

Since his death there have been countless articles written about Eddie Hubbard's exploits as a flyer, air mail pilot and airline builder. U.S. West Communications has a wall carving commemorating aviation, on its building containing switching equipment at 2nd and Lenora, Seattle. The building was built in 1986 and the City of Seattle master use permit required a retail storefront and a courtyard or some other type of street-level enhancement. U.S. West or at that time Pacific North-west Bell proposed a mural, an idea the City accepted. The artist was Mara Smith who has done artwork at other locations in the downtown area. This interesting work in brick shows Bill Boeing, Eddie Hubbard and the CL-4S seaplane that carried the first North American International air mail from Vancouver, British Columbia to Seattle. Another Seattle tribute is on the walls of Rainer Square. A series of photographers of the early Boeing days included one of Hubbard and Boeing leaving Union Lake, Seattle in the CL-4S for Vancouver, B.C. It also features the 40-B mail plane which Hubbard was instrumental in having the Boeing Airplane Company mass produce.

Artist Mara Smith's brick sculpture of Eddie Hubbard and William E. Boeing on U.S. West Building, 2nd and Lenora, Seattle.

In 1988 at the seventh annual Pathfinder Awards program in Seattle Eddie Hubbard was added to the list of people such as Bill Boeing, Phil Johnson, Lou Marsh, Glenn Martin and Claire Egtvedt to name a few. The Museum of Flight has created the Pathfinder Award program to recognize those individuals whose vision and efforts furthered the science of aviation in the Northwest.

75th anniversary first flight cover created by the author.

Chronology of Events in the Career of Eddie Hubbard

1907 Arrived in Seattle from San Francisco at age 18.

1915 **November:** Was the first pupil in the Aviation School of the Northwest and first in the Northwest to obtain an aviator's license from the Aero Club of America.

1916 **June:** He prepared a learned article on aviation for the *Salt Lake Tribune.*

August: Flew delegates from Red Cross convention over Lake Washington.

December: Joined Pacific Aero Products Company - which became Boeing Airplane Company, May 1917.

1917 **August:** Left Boeing to become Army flying instructor, Rockwell Field, San Diego.

1918 **May:** Returned to Boeing after putting on great aerobatic show at San Diego.

1919 **February 17:** He and Boeing attempted flight to Vancouver, B.C. Aircraft damaged at Anacortes.

February 27: They successfully reached Vancouver and Eddie did a series of stunts for the local citizens. Returned to Seattle with the first North American International air mail.

April: Flew the first Senator to keep an appointment by airplane, Miles Poindexter, from Seattle to Bremerton.
April: Put on a show over Seattle for returning soldiers of the 91st Division.
May: Flew first passengers of Boeing Aerial Taxi around Seattle.
May: Met two Canadian flyers and guided them to landing spot at Jefferson Municipal Golf course.
May: Dropped flowers on University of Washington to commemorate 51 students who gave their lives in WWI.
June: Did stunts for "Big Brothers" picnic.
June: Flew Bill Boeing to Victoria, B. C., for luncheon address to Aerial League of Canada and Rotary Club.
June: Flew Boeing officials over Seattle.
June: Flew second honeymoon couple from Seattle to Tacoma.
July: Put on a flying exhibition for Eddie Rickenbacker over open air banquet.
July: Flew Red Cross official from Seattle to Port Townsend to catch transpacific liner.
July: Took *P.I.* newspaper reporter aloft to do item on a parade.
August: Newspaper photos of Eddie flying mother and daughter over the city.
August: Took State Senator Lamping up for a survey flight for a municipal landing site.
September: Delivered tires by air to Everett. Tire company had ad in tire magazine about the feat.
September: Took cameraman up to photograph new subdivisions for real estate company.
September: Took cameraman up to photograph arrival of fleet.
September: Delivered eyeglasses to passenger bound for Orient at Port Townsend by dropping well-wrapped package on deck of ship.
November: Flew a senior Miss from University of Washington over stadium to drop football.

December: Flew Sir Arthur Whitten-Brown of Alcock and Brown fame over city in new Boeing plane.
December: Flew a washing machine from Seattle to Tacoma strapped to wing of plane.

1920 **January:** Famous singer Tetrazzini was to fly over city. Manager nixed as cold air would hurt her throat. Eddie took lady reporter up instead. She did glowing write-up on trip.
March: Newspaper reports of Eddie flying ore from interior B.C. mine to ocean.
April: Boeing and Eddie go to San Francisco aviation exhibition with 3 planes and a sea sled.
May: Many newspaper ads during past year for Aerial Taxi.
May: Newspaper picture and write-up of Eddie and new car he had just purchased.
June: Spoke to Board of County Commissioners along with Bill Boeing about a badly needed landing field site.
June: Performed stunts for boy's picnic sponsored by the Elks Club.
June: Flew Boeing sales manager Herb Berg from Lake Union to his home near Tacoma.
June: Flew Commissioner to Sand Point air field dedication ceremony.
July: Delivered a BB-1 flying boat to customer in Vancouver, B.C.
August: Flew to Victoria to pick up honeymoon couple for trip to Seattle.
August: Flew to Vancouver to give testimony before Canadian Air Board about BB-1 crash.
August: Flew Bill Boeing and party from Campbell River to Buttle Lake trout fishing.
September: Dropped theatre passes over Seattle.
September: Rushed printing molds from Seattle to *Victoria Daily Times* so next edition could be printed.
September: Flew Chief of Engineering Division of Air Service over Sand Point airfield site.

October: Postmaster General announces Seattle-Victoria air mail contract let to Eddie.
October: Victoria's two and Seattle's three newspapers ran many articles and pictures about the new Foreign Air Mail Route between Seattle and Victoria.
October: Flew two passengers from Seattle to Victoria.
October: Late leaving Victoria on third trip and flew in the dark.
October: Engine problems on fourth flight and stopped over night at Port Ludlow.
October: Flew Seattle Postmaster Edgar Battle over air field site at Sand Point.
November: Coming back from Victoria, ran out of gas. Landed in ocean off Seattle. After drifting and shouting for help he finally attracted attention. Someone came out with gasoline, allowing Eddie to proceed to Lake Union.
December: Fought 90 mile per hour gale to make air mail flight.

1921 **February:** Future of air mail service hangs in balance awaiting Congress to approve funds.
April: Continuation of Seattle-Victoria air mail approved.
July: Takes wife and friends on a tour of Vancouver Island and Gulf Islands.
August: Made midnight mail run Victoria to Seattle.
August: Two newspaper reports on his mail pick-ups.
September: Another report on his mail pick-ups.
September: *Victoria Times* had full page report including photos on the success of the Seattle-Victoria air mail. Over one million letters carried in ten months.
November: Forced down in blizzard in Puget Sound after hunting trip. Plane badly damaged and his friend drowned in an attempt to swim for help.

1922 **March:** Rushes picture of French Marshall Joffre from Victoria to Seattle for *P.I.* Great attention given to same-day picture from Canada in newspaper.

March: Flew Seattle sheriff and bloodhounds to Olympic Peninsula to hunt bank robbers and killers.

1923 **January:** In twenty-eight month period flew fifty tons of mail.

March: Crashed off Victoria without serious injury. All mail recovered and B-1 shipped to Seattle for rebuild.

April: Full page weekend supplement in *Seattle Times* on Seattle's aerial postman.

June: He lost contract to Alaska Airways, however their larger plane was much more expensive to operate and they contracted Eddie to take mail in his efficient B-1.

1924 **April:** *Victoria Daily Colonist* pushing for air service between Victoria and Vancouver, citing Hubbard's success.

May: Boeing President Gott promotes possibilities of mail and passengers to Alaska pointing out Hubbard's success.

1925 **June:** He crashes into Lake Washington testing Boeing aircraft and not injured.

July: Made test flight in new Boeing aircraft to San Francisco and return.

October: Flew man to Victoria to catch up to his wife and child.

November: Quoted in Victoria paper about air mail contracts from Seattle to Los Angeles.

1926 **March:** Newspaper item that mail will soon be able to go from Victoria to Seattle via Hubbard then to Pasco by train and on to New York by air.

October: Plane substituting for Eddie crashes in Victoria. Mail recovered and pilot not seriously hurt.

1927 **January:** Hubbard did submission and new company, Boeing Air Transport, won contract for air mail between Chicago and San Francisco.

January: Hubbard suggested Boeing produce twenty-five 40-B mail and passenger planes for the new venture.

March: Salt Lake City newspaper have many write-ups on Hubbard as Vice President and Operations Manager of Boeing Air Transport headquartered in that city.

March: He addresses Salt Lake City Chamber of Commerce luncheon about pluses of new air mail contract with passengers service which was not available when Post Office flew the mail.
March: Editorial in *Salt Lake Tribune* lauding Hubbard and Boeing Air Transport.
April: Seattle *P.I.* had big write-up and photos of Boeing Hubbard and Gorst announcing Pacific Air Transport Company's passenger and mail service between Seattle and Los Angeles.
June: Gives up Victoria-Seattle air mail route.

1928 **July:** At Bill Boeing's request he purchases new landing field in Reno.
November: First flight of new Boeing Model 80 trimotor twelve passenger airplane into Reno to open new field. Bill Boeing christens it Hubbard Field.
November: He praises Salt Lake City for improvements to the airport.
November: Passenger in a Boeing plane that has a minor accident at Granger, Wyoming. No one injured.
December: He suddenly passes away in the hospital after a stomach operation at the young age of 39.
December: His death was front page news across the United States and many parts of Canada.
December: Seattle, Victoria and Salt Lake City newspapers had glowing editorials about this outstanding man.
December: He is buried in Seattle.

1929 **January:** When his will is probated his widow Mildred receives $1,200,000.
His B-1 flying boat, the only one built, hangs in the Museum of History and Industry in Seattle.

Since his death their have been innumerable articles in newspapers, magazines and books about one of the world's great pioneer aviators and airline builders. In a prophetic early statement given in Victoria, Eddie Hubbard said, "I want to prove that aeroplanes can carry mail in regular and efficient service and I want to show that carriage of mails by air can be made a commercial success."

THE B-1

The Boeing B-1 was the first commercial aircraft designed, engineered and built by the Boeing Airplane Company. The hull was made of laminated wood veneer and the wings were framed with Washington spruce.

There was only one B-1 flying boat manufactured. Its first test flight was on July 12, 1919, conducted by Eddie Hubbard, who flew to an altitude of only 500 feet. The B-1's hull was designed by George Pocock, a well-known Seattle rowing shell designer and builder.

In a July 28, 1919 memorandum to Boeing at Ketchikan, Alaska, Edgar Gott, General Manager, said they were having air pump problems with the engine. However a representative of the manufacturer Hall-Scott was at the plant and the pump was satisfactorily redesigned. Gott also mentioned a J.A. Chilberg was interested in purchasing the B-1.

On Sunday, August 3rd Hubbard took the B-1 on a test flight with the redesigned air pump. Hubbard mentioned to Gott that he would rather take off from Lake Washington and arrangements were made to tow the aircraft to Madison Park. In an August 4th memorandum to Boeing at Ketchikan, Gott said,

> "You will no doubt be gratified to know that the B-1 made another flight and barring a few small details performed exceptionally well. The new air pump put on the motor would have functioned very well had it not been for the pieces of shellac on the valves, which caused a little

trouble. This has been remedied and we expect, when we make the next flight, which we hope to do this afternoon, that the trouble will have all disappeared.

The ship left the water with two passengers and about 20 gallons of gasoline with astonishing rapidity and climbed better than 1,000 feet in the first two minutes. The maneuverability of the ship is now all that could be required.

Mr. Straub, of the Hall-Scott Company, was on hand at the flight and noted the difficulty and danger experienced in starting the motor. He is, and has been, in touch with his people in San Francisco with the view of taking care of this feature by the installation of an electric starter. The radiating surfaces do not seem to be quite ample enough, but the ship was not in the air long enough to confirm this suspicion."

Claire Egtvedt, Chief Engineer on the B-1 design, later when Chairman of the Board said, "I remember the day Eddie Hubbard took it up for the first time one of the foremen offered to bet Bill Boeing it would never get off the water. Louie Marsh, Assistant Chief Engineer, and I went along with Eddie and the plane did get off the water although it experienced fuel pump trouble when it was airborne. We skimmed over the tops of some anchored boats in a series of upward swoops and dives. I understand the crowd watching from the Lake Union shore got pretty nervous before we landed, but probably not half as nervous as we were."

Apart from the first test flight the Boeing Airplane Company Aviation Log Book has the following entries:

- December 27, 1919 - 3 p.m. - flew Egtvedt and Marsh for testing of new engine and no noticeable improvement.
- December 29, 1919 - 3:10 p.m. - same 3 flew to a ceiling of 13,400 feet in 69 minutes.
- June 18, 1920 - 1:20 p.m. - Flew to Bellingham with 2 passengers - W. McCrabbe & B. Wall.
- June 29, " - 2:30 p.m. - Flew 2 passengers to Angle Lake & return.
- June 29, " - 5:30 p.m. - Flew Mr. Boeing around sound.
- June 30, " - 8:50 p.m. - Moonlight trip for Mr. Boeing.

- July 3, " - 9:00 a.m. - Trip to Vancouver with M. E. Boeing & Mr. & Mrs. Keena.
- July 4, " - 4:30 p.m. - Return from Vancouver with Mr. & Mrs. Keena.
- July 6, " - 6:30 a.m. - Trip to Victoria with Mrs. Brown & Jessica Ross.
- July 9, " - 7:00 a.m. - To Vancouver & return with Mr. Boeing.
- July 24, " - 2:30 p.m. - Photos of Seattle with Pierson - Strang.
- August 11, " - 5:30 p.m. - Trip to Bremerton & Crystal Springs with 3 passengers.
- August 14, " - 2:30 p.m. - To Victoria with Hubble.
- August 14, " - 5:00 p.m. - Brought bridal couple from Victoria.
- August 14, " - 6:00 p.m. - Trip to Crystal Springs.
- August 24, " - 8:00 a.m. - Trip to Campbell River, B.C.
- August 25, " - 9:15 a.m. - Fishing trip to Buttle Lake with Mr. Boeing, J.T. Keena & J.R. Rithet.
- August 26, " - 10:50 a.m.- Return from Campbell River.

On the Seattle-Victoria air mail route plus taxi and pleasure flights the B-1 wore out 6 Liberty engines in flying 150,000 miles.

Les Hubbel and Mike Pavone were the mechanics looking after the B-1. Les said, "Considering the lack of instruments and difficulty of route Hubbard had very few accidents. Now and again the hull would be damaged by logs. Once, flying a few feet off the water in fog, he got too close to the surface. A wave hit the hull and opened it. When he landed in Lake Union the plane promptly sank. But we pulled it to the surface, patched it up, and it went on flying as well as before." Mike said, "They had 3 Liberty engines with one in use and two spares. When an engine was changed I would go for a test flight with Eddie. The B-1 was originally powered with a 200 hp Hall-Scott 6 cylinder engine but was soon replaced with the 400 hp V-12 Liberty, giving a cruising speed of 90 miles an hour."

During its flying career the B-1 had no less than five registrations. In 1920 the United States was not a member of the International Aeronautical Convention of 1919 which founded identification and licensing of aircraft. With the B-1 operating in Canada the authorities insisted it carry identification and "G-CADS" was allotted. Canada used the "G" until 1929 when "G" was assigned solely to Great Britain. As the aircraft was not Canadian-owned and the prefix "N" had been reserved for the United States, the first identification painted on the B-1 was "N-CADS". In 1923 the American National Board of Underwriters started an unofficial licensing system for aircraft it insured and the B-1 was registered as "N-ABNA." The "N-ABNA" registration was used longer than any other. In 1927 the U.S. adopted a licensing system with registration starting with "N" followed by "C" for commercial; however, the B-1 did not qualify under the new air worthiness requirements and was given the number 4985. Soon after it was able to qualify under a grandfather clause. The B-1 became "NC-1974." When Hubbard rejoined Boeing in 1927 to become Vice President of Boeing Air Transport as Operation Manager in Salt Lake City he sold the B-1 and it continued on the Victoria Air Mail run out of Renton until 1928. It was then stored outside at the Renton airport, formerly known as Bryn Mawr Field. It was later stored at Boeing Field. A great deal of deterioration set in and in the early 1950's some of Boeing's oldest employees restored the battered "one of a kind."

The B-1 was purchased from Northwestern Air Service by the Seattle Historical Society. The following is an extract from Society minutes of a meeting on March 10, 1942 at the home of Mrs. Thomas Green with twenty-three members in attendance: "The Boeing Company gave us the opportunity of purchasing the First Air Mail plane (B-1) that was piloted by Eddie Hubbard in early days. It is crated and the Boeing Company will care for it until we have a home ready for it. Mrs. Frink moved, Mrs. Sanford seconded we purchase the plane at the price offered ($125.00) one hundred twenty-five dollars. Motion carried." The minutes of the May 12th meeting corrected the March 10th minutes to state that the price of the Hubbard plane was two hundred dollars. The check was made out to Northwestern Air Service.

It remained in storage until 1951 when it was renovated and reassembled by Boeing employees at Renton under the supervision of Nick Carter. The B-1 now hangs in the Philip Johnson (Boeing President 1926-1933, 1939-1944) Memorial Aviation Wing at the Museum of History and Industry near the University of Washington.

ORIGINAL SPECIFICATIONS AND PERFORMANCE OF THE B-1

Span	50 ft. 3 in.
Length	31 ft. 3 in.
Wing Area	492 sq. ft.
Powerplant	200 hp Hall-Scott L-6.
Empty Weight	2,400 lbs.
Gross Weight	3,850 lbs.
High Speed	90 mph.
Cruising Speed	80 mph.
Service Ceiling	13,300 ft.

B-1 was originally registered in Canada as G-CADS. The aircraft was American-owned and the prefix "N" had been reserved for the United States. The first registration painted on the B-1 was "N-CADS."

In 1923 the American National Board of Underwriters started an unofficial numbering system for aircraft it insured. The B-1 registration was changed to "N-ABNA." This registration was used longer than others in the flying life of the B-1.

B-1 in a sad state of disrepair. Being dismantled for restoration.

Restoration of B-1 by Boeing employees and retirees.

Restored B-1 admired by military personnel in front of a C-97A Stratofreighter at the Boeing Renton plant.

Old and new bathing suits pose with the first commercial aircraft built by the Boeing Airplane Company.

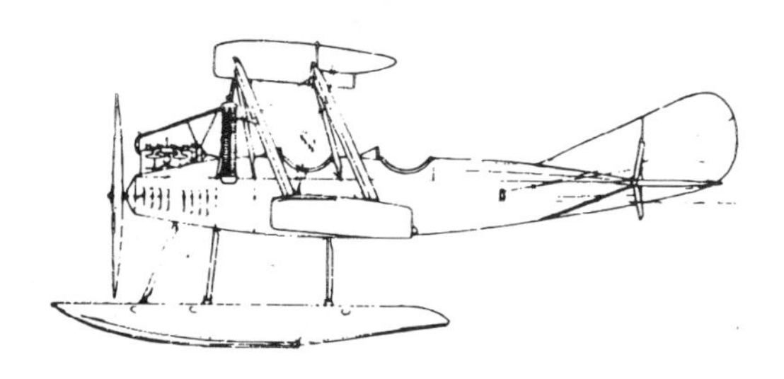

CL-4S

Originally built as one of fifty "C" type training seaplanes for the United States Navy and given the designation C-650-699. An additional aircraft was built for Bill Boeing's personal use and was designated C-700.

The C-700 was modified in December 1918 and redesignated CL-4S. A new Hall Scott L-4 engine replaced the Curtiss along with minor aerodynamic changes.

The CL-4S made two important first air mail flights. On March 3,1919 Bill Boeing and Eddie Hubbard flew the first North American International air mail from Vancouver, British Columbia to Seattle, Washington. Sixty envelopes were carried. The last one auctioned by Charles G. Firby two years ago went for almost $4,000!

The second air mail flight was by Eddie Hubbard, October 15,1920, on the first North American contract air mail route from Seattle to Victoria, British Columbia and return to Seattle. Although this route was designated United States Foreign Air Mail Contract No.2 (FAM2), it was in fact No.1. It actually started fifteen days ahead of all other U.S. Foreign Air Mail Contracts. First flight air mail covers are valued today at $200.

Hubbard used the CL-4S on his 10 monthly mail trips to Victoria for several months before purchasing the B-1 flying boat.

Eddie Hubbard leaving Lake Union hangar with a passenger in Bill Boeing's personal seaplane, the CL-4S.

MODEL 40

The Model 40 prototype with a Liberty 400 hp engine was built for the Post Office and first flown in 1925. The project was shelved until 1927 when Eddie Hubbard made two suggestions: produce an improved Model 40 with an all-metal body and upgrade to a Pratt & Whitney 420 hp Wasp air-cooled engine. This was designated as the 40A. The air-cooled Wasp engine was not only more powerful but was 700 pounds lighter than the water-cooled Liberty. The air-cooled engine didn't need the plumbing required for a water-cooled engine. This made it possible to carry a bigger payload of mail and express. With the new engine the 40A's had a full load speed of 129 mph, a landing speed of 57 mph. a rate of climb of 600 feet per minute and a 12,000-foot ceiling.

Boeing Model 40-C, which had two cabins and carried four passengers vs two passengers carried in the 40-A. This aircraft was used on CAM8 United States Contract Air Mail Route No. 8 from Seattle to Los Angeles by Pacific Air Transport Inc., commencing September 15, 1926. The Boeing Airplane Company purchased a substantial interest in Pacific in 1928, which added "Boeing Lines" to its logo. In front of the cabin is the American Railway Express label.

The 40-B4 had two cabins to accomodate four passengers. By 1931 the logo was now United Air Lines plus Pacific Air Transport and Boeing for United States Contract Air Mail Route 8 between Seattle and Los Angeles.

There were many other improvements over the original '40,' such as the landing gear. It used the new Boeing Oleo shock absorbers, which made landings and take-offs on rough fields a lot smoother. The wings of this biplane were spruce spars and ribs, fabric covered, braced with a new type of wire and designed with a new airfoil section that gave performance characteristics considered excellent for its time. Under each wing was a 250,000 candle power landing light and two flares for night forced landing. These flares would light up almost one square mile of ground. Forced heat was provided for the two passenger cabin along with a dome light for night travel. Two compartments accommodated 1,200 pounds of mail. There was a door on each side of the body with safety glass windows which could be opened for ventilation. The windows were large enough so that passengers got a good view of the passing scenery. The cockpit was open and behind the passenger cabin and the tail skid was steerable.

Twenty-five 40A's were built. Boeing Air Transport took twenty-four and Pratt & Whitney bought one for engine testing.

In 1928 the Model 40A's of Boeing Air Transport were returned to the Seattle plant for engine upgrading from the 420 hp Wasp to the improved 525 hp. Pratt & Whitney Hornet. The Boeing Air-

plane Company had a supply of the new Pratt & Whitney 525 hp Wasp engines at the Seattle plant but they were for use in Navy planes under construction. Bill Boeing received permission from Pratt & Whitney to use them in the 40A's. The engine manufacturer satisfied the Navy that by the time the engines were needed for their new planes a new supply would be at the Boeing factory. After the engine replacement the designation was 40B.

In 1929 a new Model 40B-4 was produced which was able to carry four passengers instead of two. Thirty-eight were manufactured.

Soon after Boeing took over Pacific Air Transport in January 1927 a fleet of ten 40C's replaced the aircraft Gorst had been using. The 40C was similar to the 40A but carried four passengers.

The executive 40-B built for the Standard Oil Company of California (Chevron) designated 40Y by Boeing and No. 2 by Socal.

Two executive 40's were built. One designated 40X for Associated Oil Company and 40Y for the Standard Oil Company of California.

Canadian built Model 40-B designated as 40-H. Henry Hoffar was president of Boeing Canada so Bill Boeing made the Canadian production H's instead of B's. This is first air mail flight from Fort St. John to Fort Nelson, B. C. Pilot is Sheldon Luck in the middle with the wool face mask for the January below-zero weather. The Model 40 had an open cockpit.

Five 40H-4's were built by Boeing Aircraft of Canada Ltd. in Vancouver, B. C. The Vancouver plant was at the former Hoffar Shipyards. Henry Hoffar had built aircraft at this location and became President of the Boeing Canadian subsidiary. Boeing used the initial "H" for Hoffar to designate the Canadian models. Two were sold to New Zealand Airways and both ended up in New Guinea. One crashed in 1939 and the other was destroyed by enemy action in 1942. On a cold clear day in January, 1937 CF-AMQ carried the first air mail from Fort St. John to Fort Nelson, Gold Bar, Finlay Forks and Fort Grahame in Northern British Columbia. The pilot was Sheldon Luck, flying for Grant McConachie's United Air Transport Limited. McConachie later became President of Canadian Pacific Airlines. In the picture Sheldon is in the middle with Canadian Post Office Inspector George Beatty on his right and Engineer Red Rose on his left. Notice that Sheldon has a wool face mask to give him some protection from the cold in the open cockpit. The mail bags are on the ground between the pilot and his engineer.

The first catalog produced by the Boeing Airplane Company in the fall of 1928 promoting its aircraft had pictures of the Model 40 and said,

"Twenty-five mail planes, built by Boeing and operated by the Boeing Air Transport, Inc., are operating safely and speedily on day and night schedules from the Pacific to the Great Lakes ... another enviable record and another tribute to the dependability of Boeing airplanes. This is the longest air mail route in the world operating on schedule and the planes now have two million miles of safe operation behind them. Forty-six per cent of the total time flown on the transcontinental route calls for night operation—the success of which carries further proof of reliability.

"Five four-passenger Boeing planes are now being operated on the Seattle-Los Angeles air mail route by the Pacific Air Transport, Inc. Part of this line is also operated on night schedule.

"We offer to the air mail operator our Model 40, Series A, B and C, all of which carry the Department of Commerce Approved Type Certificate. These planes are designed for both mail and passenger operation. Cabins are metal lined with sound-proofed walls and ceiling. The seats are deeply upholstered and comfortable. Space is provided for luggage at the forward end of the cabin and under the rear seat. Dome lights provide night illumination and every plane is equipped with forced heating and ventilation. Windows of non-shatterable glass are of ample size to afford an excellent view."

The catalog also included the following data with a full load:

PERFORMANCE

Model 40-A	Model 40-B	Model 40-C
High Speed		
129 mph	129 mph	132 mph
Landing Speed		
57 mph	57 mph	57 mph
Rate of Climb		
600 ft (per min)	600 ft (per min)	800 ft (per min)
Climb in 10 minutes		
4,890 ft	4,890 ft	6,320 ft
Service Ceiling		
12,000 ft	12,000 ft	15,100 ft

Model 40-A	Model 40-B	Model 40-C
Take-off Run		
655 ft	655 ft	350 ft
Take-off Time		
12 seconds	14 seconds	14 seconds
Wing Loading		
11.15 lbs/sq ft	11.0 lbs/sq ft	11.15 lbs/sq ft
Power Loading		
14.84 lbs/hp	14.63 lbs/hp	11.58 lbs/hp
Power Plant (Pratt & Whitney)		
"Wasp"400 hp	"Hornet"500 hp	"Wasp" 400 hp
	WEIGHT	
Weight, empty		
3,522 pounds	3,313 pounds	3,543 pounds
Pilot		
180 pounds	180 pounds	180 pounds
Fuel (140 gallons)		
840 pounds	840 pounds	840 pounds
Oil (10.5 gallons)		
88 pounds	88 pounds	88 pounds
Pay Load		
1,579 pounds	1,428 pounds	1,445 pounds
Gross Weight, loaded		
6,000 pounds	6,079 pounds	6,075 pounds
	DIMENSIONS	
Wing Span		44 feet 4 1/4 inches
Length overall		33 feet 4 7/16 inches
Height overall		11 feet 8 1/2 inches

A total of eighty-one 40's were built. The aircraft was not only an economic success but was the first mass production commercial plane manufactured by the Boeing Airplane Company. This dramatic entrance into the commercial field of aviation was the innovative idea of Eddie Hubbard.

Two 40's have survived. One hangs in the Rosenwald Museum in Chicago and the other is at the Henry Ford Museum in Dearborn. It is unfortunate a third is not available to hang in the Museum of Flight in Seattle where it all happened.

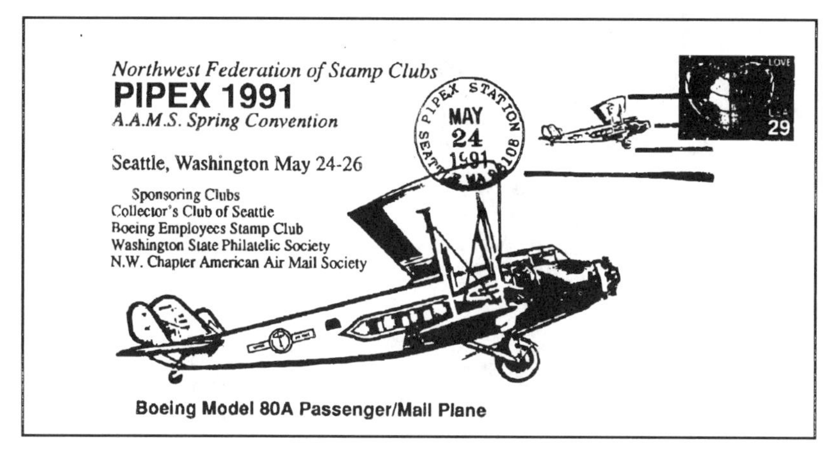

Boeing Model 80A Passenger/Mail Plane

MODEL 80

Business for Boeing Air Transport was booming. Mail poundage and the number of passengers were on the increase. As a result Boeing Air Transport required larger aircraft for the San Francisco-Chicago route. Phil Johnson, Boeing President had such a project in mind. In 1928 the Boeing Airplane Company produced a tri-motor twelve passenger aircraft not only for use on the San Francisco-Chicago route but also for sale to other airlines.

For its time it was a very luxurious machine with leather upholstered seats that reclined, heated cabin, forced air ventilation, hot and cold running water, lavatory, insulated and sound proof cabin, reading lamps, box lunches served, and the world's first stewardesses. To qualify as a stewardess the women had to be registered nurses. The plane was aptly called, "the pullman of the air" with a speed 128 m.p.h. The pilot and co-pilot had the benefit of a two-way radio and an enclosed front cockpit.

In the initial production four were built for Boeing Air Transport. Later models were the 80A which carried eighteen passengers. There was an 80B-1 which had an open cockpit. After the seasoned mail pilots that were used to the open cockpit of the Model 40's got used to the comfort of an enclosed cockpit no more "open-air" models were built.

Boeing President Phil Johnson believed air traffic would rapidly increase and require larger planes. He pushed to replace the four-passenger Model 40's with the twelve-passenger Model 80. A later version carried eighteen. Note the Boeing Air Transport logo.

Custom-built Model 80 for Standard Oil Company of California (Chevron).

Original Boeing Airplane Company building. Often referred to as the "Red Barn," it now is part of the Museum of Flight at Boeing Field, Seattle.

Bibliography

BOOKS

Aircraft Year Book, Doubleday, Page and Co., 1920.

Aircraft Year Book, Small, Maynard & Co., 1921.

Bauer, Eugene E., *Boeing in Peace and War,* TABA Publishing, Enumclaw, Washington, 1990.

Bowers, Peter M, *Boeing Aircraft Since 1916,* Naval Institute Press, 1993.

Davies, R.E.G. *Airlines of the United States Since 1914,* Smithsonian Institution Press,1965.

Mansfield, Harold, *Vision,* Duell, Sloan and Pearce, New York, 1956.

Munson, Kenneth, *Airliners from 1919 to the Present Day,* Peerage Books, London, England, 1972.

Oliver E.Allen, *The Airline Builders,* Time-Life Books, Virginia, 1981.

Redding, Robert & Yenne, Bill, *Boeing - Planemaker to the World,* Crown Publishers, Inc., 1983.

Rolfe, D. & Danydoff, A., *Airplanes of the World,* Simon and Schuster, New York, 1962.

Satterfield, Archie, *Alaska Bush Pilots in The Float Country,* Bonanza Books, New York, 1975.

Sieh, Pingwen & Blackburn, Lewis J., *Postage Rates of China 1867-1980,* Directotate General of Posts, Taipei, Taiwan.

The Boeing Company, *Pedigree of Champions Boeing Since 1916,* Fifth Edition, 1984.

Gunston, Bill, *The Illustrated Encyclopedia of Propeller Aircraft,* Exeter Books, New York, 1980.

PUBLISHED DOCUMENTS AND BULLETINS

The Postal Bulletin, United States Post Office, October 5,1920, No.12376.

The Postal Bulletin, United States Post Office, June 29,1937, No.17216.

Contract For Foreign Mail Service By Aircraft, United States Post Office, October 15,1920 to June 30,1921.
Contract For Foreign Mail Service By Aircraft, United States Post Office, July 1,1922 to June 30,1923.
Contract for Foreign Mail Service By Aircraft, United States Post Office, July 1,1925 to June 30,1926.
Foreign Air Mail Contract F.A.M. Route 2, Seattle to Victoria, Seattle-Victoria Air Mail Inc., June 30,1933 for four years.
Information Bulletin No.29 from George S. Wheat Vice President, United Aircraft & Transport Corporation to Executives of United Airlines and subsidiary companies.

NEWSPAPERS and PERIODICALS

Aero Digest, March, 1927.
Air Pictorial, November, 1966.
American Reveille, Bellingham, WA., 1920.
Anacortes American, Anacortes, WA., 1919.
Boeing Airliner, July, 1978.
Boeing News, July, 1936; August, 1937; November, 1937.
BNA Topics, July-August, 1958.
Chronicle, Spokane, WA., 1920.
Duwamish News, Seattle WA., 1920.
Grade Club Magazine, Seattle, WA., June, 1927
Hardware Age, 1919.
Herald, Bellingham, WA., 1920.
Herald, Port Angeles, WA., 1920.
Hoquiam Washingtonian, Hoquiam, WA., 1920.
Jack Knight Air Log, Spring 1959.
London Philatelist, Vol 80, January, 1971.
Nevada State Journal, Reno, NV., 1928
Port Townsend Daily Leader, Port Townsend, WA., 1920.
Record, Ferndale, WA., 1920.
Reno Evening Gazette, Reno, NV., 1928.
Salt Lake City (Utah) *Desert News,* 1927.
San Diego Union, San Diego, CA., 1918.
Salt Lake City (Utah) *Telegram,* 1927.
Spirit, Seattle Chamber of Commerce Bulletin, 1920.

Tacoma Times, Tacoma, WA., 1920.
The Airpost Journal, March, 1958, October, 1959. March 1984.
The American Philatelist, July, 1985.
The Comox Argus, Courtenay, B.C., 1920.
The Daily Colonist, Victoria, B.C., 1919-1928, 1937, 1949, 1958.
The Daily Province, Vancouver, B.C., 1919-1920.
The Salt Lake Tribune, Salt Lake City, 1919, 1927-1928.
The Seattle Daily Times, Seattle, WA., 1915-1928, 1941, 1942, 1989.
The Post-Intelligencer & Seattle Post-Intelligencer, Seattle, WA, 1915-1928, 1938, 1950, 1951,1957,1958,1960, 1976, 1989.
The Seattle Star, Seattle, WA., 1920-1921, 1928.
The Tacoma Sunday Ledger Tribune, Tacoma, WA, 1919, 1920, 1951, 1970.
Tribune, San Diego, CA., 1920.
Vancouver Daily Sun, Vancouver, B.C., 1919-1920.
Victoria Daily Times, Victoria, B.C., 1919-1928, 1937,1942.
Vancouver Morning Star, Vancouver, B.C., 1926.
Vancouver World, Vancouver, B.C., 1920.

CORRESPONDENCE AND INTERVIEWS

Airport Authority of Washoe County, Reno, NV., correspondence, November 13,1990; December 28,1990; February 19,1992; March 11,1992.
Armstrong, Lynne R., correspondence, October 27 & November 7,1992.
Barnes, Harry Jr., interview, Victoria, B.C., April 11,1991.
Benson, George, E, correspondence, November 23,1993.
Blumenthal, Frank H., correspondence, May 18 & June 2,1988.
Boeing, William E. Jr., interview, Seattle, WA., Jan. 20 & Feb. 12, 1996.
Brewis, Greg, correspondence, July 14, 1990, October 12,1991.
Burrell, Basil correspondence, February 25 & March 12,1988.
Carr, Harold, correspondence, December 20, 1993.

Carvalho-Maia Graham, Suzzane, correspondence, November 29, 1888; December 3,1988; December 19,1988; July 5,1990; October 4,1990; March 17,1992; March 30,1992; June 17,1992; interview, June 12,1993, June 15,1993; November 4,1993; February 7,1994; March 4,1994; March 16,1994; interview, April 30, 1994, May 18, 1994.
Cressy, Ted H., interview, March 14, 1987, January 10,1988.
Chicago Historical Society, correspondence, October 14, 1995.
Crosby, Gail & Ann Scott interview, October 26,1995
Davies, R.E.G., correspondence, March 4, 1988, March 14, 1988.
Dube, Timothy, correspondence, February 17, 1993, March 1,1993.
Firby, Charles correspondence, April 12 & April 20,1995.
Fish, Harriet U. correspondence, September 26, November 20 & 24, 1995, January 6 & 15, 1996.
Frink, Frank G., correspondence, April 14,1993.
Frink, Margaret B., correspondence, April 26, 1993.
Furlow, Elizabeth, correspondence, May 7, August 25, 1995.
Grigore, Julius, correspondence, October 18 & 27, November 9, 1995.
Hillman, Thomas A, correspondence, November 29,1988.
Johnson, Ken R., correspondence, November 8 & November 23,1989
Leary, William M., correspondence, October 29, November 11, December 5,1987.
leBeau, David, Chevron, correspondence, August 10, September 19,1995.
Malott, R.K. (Major,Ret'd), correspondence, July, 1985.
Mortensen, Marjorie Lee, Librarian Nevada Historical Society, Reno, NV., correspondence, November 14,1990.
Moroney, Rita L., correspondence, August 10,1988, October 4,1988.
Morrow, Ms. Lory, correspondence, May 4,1994.
Nicastro, Kathie. O., correspondence, October 25,1988.
Nortum, Rose, correspondence, April 16, 1992, March 7,1993.
Nute, Brud, interview, June 11,1992, Bainbridge Island, WA., correspondence, April 13,1993.

O'Neill, Jane, correspondence, November 3 & 9, December 7, 1995.
Parrish, Sandra, correspondence, April 24,1992.
Pfouts, Betty J., correspondence, October 22, 1991, January 23, January 29, February 4,1992.
Plomish, Walter, correspondence, August 5 & September 17,1989.
Pratt & Whitney, Canada correspondence, April 27, 1995.
Upper, Gloria Huntington, correspondence, July 7,1993.
Ridout, Cec., interview, Victoria, B.C., April 12,1991.
Rogers, Michael correspondence, March 14,1988.
Rutledge, Ann, correspondence, March 23,1990.
Salt Lake Tribune, correspondence, June 18 & July 5,1990.
Sioras, George, correspondence, June 8,1988.
Short, Murrieal, correspondence, May 31, June 12, June 20, July 24, August 27, November 10, 1995.
State Historical Society of Iowa, correspondence, September 7 & 25, 1995.
U. S. West Communications, correspondence, May 9, June 26, 1995.

PUBLISHED DOCUMENTS

Canadian Air Force Court of Inquiry Report, August 20,1920.

Index

A

Admiral Line 114
Aerial League of Canada 49, 52, 59-60
Aerial Taxi 11, 44-48, 56, 63-64, 69, 71-72, 75-76, 78, 80-81, 83-84, 97, 125, 185
Aero Club of America 16, 19, 184
Aero Club of Canada 38
Aero Club of the Northwest 129
Aero Marine 114, 117, 118
Air Mail 9, 11, 12, 15-17, 31, 34-38, 41, 49, 53, 87-89, 95, 97, 99-100, 112, 114-116, 119, 121-122, 124-125, 129-130, 132, 149-150, 152, 154-155, 157, 159, 162-163, 165, 167, 169, 172, 174, 182-184, 187-188, 193-194, 199, 201, 202, 204-205, 212
Aircraft Manufacturing Company 91
Alaska 38, 42, 44, 113-115, 145, 188, 191
Alaska Airways 113-115, 145, 188
Alcock, John 129
Alcock & Brown 60, 129, 186
American Air Mail Society 107
American Legion 80
American National Board of Underwriters 194, 196
American Railway Express 164, 201-202
American Society of Mechanical Engineers 177
Anacortes, WA 32-33, 126, 184, 212
Andrews, Lt. J. C. 34
Army and Navy Club 129
Associated Oil Company 203

B

B & W 16-18, 24
B-1 64, 70, 72-75, 77-81, 83, 100, 102, 110-111, 113-114, 116, 119-120, 125, 127, 141, 143-144, 188-189, 191-199
Bainbridge Island, WA 64, 137, 179
Bandy, C. W. 137-138
Bane, Thurman H. 47
Barnes, Harry 102, 108,
Barnes, Harry Jr. 101, 108

Battle, Edgar 38, 95, 100, 122, 144, 187
BB-1 83-84, 91, 186
Beachey, Lincoln 21
Beatty, George 204
Bellingham, WA 192
Bennett, Gordon Aviation Cup 20
Berg, Herb 45-46, 66, 67, 186
Berlin, C. A. 29-30
Biffle, Ira 162
Blanche Field, Reno 167
Boeing, William E. (Bill) 9, 11, 13, 15-16, 18, 23-24, 28, 30, 32-33, 35, 38, 41, 43, 45-46, 53, 56, 59-61, 68-69, 73-74, 80, 87-88, 90, 100-102, 121-122, 125-126, 129-130, 133-134, 150, 152, 154-156, 159, 162-164, 166, 168, 172, 174, 176, 179, 181, 183-185, 189, 191-193, 195, 199-200, 203-204, 206-207, 213-214
Boeing, Mrs. 159
Bishop, Harry 102, 114, 123
Blue Funnel Line 101, 108
Boeing 314 115
Boeing Air Transport Inc. 124, 154-156, 162-169, 179, 181-182, 188, 194, 202, 205, 207-208
Boeing Aircraft of Canada Ltd. 182, 204
Boeing Airplane Company 10-11, 16-17, 24, 26-28, 31, 41-42, 45, 48, 75-77, 83, 89, 95, 116, 120-121, 125, 130, 143, 150, 154-155, 179, 181, 183-184, 191-192, 198, 201, 203, 205-206, 208-209
Boeing B-1E 118-120
Boeing Commercial Air Service 44
Boeing Field 209
Boeing Model 40 -- See Model 40
Boeing Model 80 -- See Model 80
Boise, Idaho 163
Boone, Ike 81
Bremerton, WA 26, 31, 43-44, 64, 185, 193
Bremerton Yacht Club 64
Brenton, Capt. Hibbert 91
British Columbia Gov't. Library 132
Brochie Ledge, Victoria 141
Brown, Arthur Whitten 60, 129-130
Brown, Lt. Harry 49-53
Brown, Hon. Frederick V. 73
Bryn Mawr Field 194
Burleson, Postmaster General 97

Burns, Robbie 38
Butler, Catherine 56
Buttle Lake, B. C. 73-75, 101, 134-135, 186, 193

C

C. P. R. 78, 130, 142
C-700 18, 28, 30, 32, 130, 199
Cadboro Bay, B. C. 59-60, 73, 115, 126

Campbell, C. L. 44, 46
Campbell River, B. C. 7, 73-74, 134, 136, 193
Canada 193, 195, 204
Canadian Air Board 103
Canadian Customs 59
Canadian Empress Ships 96, 98, 136
Canadian Government 33, 49
Canadian Immigration 81
Canadian Pacific Airlines 204
Canadian Pratt & Whitney Aircraft Company Ltd. 181
Cannon, Senator Howard W. 168
Carter, Nick 194
Carvalho-Maia, Suzanne 37
Cascade Motors, Seattle 64
Certificate of Flight 48
Challam County, WA 139
Chance Vought Corporation 181
Chase, Sheriff 140
Chevrolet, Louis 61
Chevron 124, 203, 208
Cheyenne, WY 159, 163
Chicago 115, 123-124, 149-150, 154, 156, 158-160, 163, 165, 167, 182, 206-207
Chicago Herald 162
Chilberg, J. A. 191
China Clipper 115
Chinese Postal Service 39
CL-4S 18, 28, 32, 35, 46, 59, 65, 71, 101-102, 120-121, 125, 130, 134, 183, 199, 200
Colley, Tom 142
Colyer, D. B. 163
Comox Argus 73
Compo, Lt. G. L. 145-146
Condon, Lt. 31

Congress 87, 112, 187
Contract Air Mail Route No. 8 201-202
Contract Air Mail Route No. 18 123, 154, 165
Coolidge, President Calvin 162
Corporate and Legal History of United Air Lines 182
County Commissioner 87-88, 90, 186
Coupeville, Whidbey Island, WA 12, 49-50, 122
Couples, Fred 52
Curtiss 146, 199
Curtiss, Glenn 20, 48-50
Curtiss Flying Boat 15, 119
Curtiss Gull 79
Curtiss HS-2L 41, 113-114
Curtiss Jenny JN-4A Canuck 48-50, 52, 59

D

Davenport, H. B. 108
Dean, Dora 72
Deeds, Charles W. 181
Delfino, Adrian 182
Depurdussin 25-27
Des Moines, Iowa 163
Detroit 136, 150, 154
Diamond Rubber Co. 66
Discovery Bay, WA 139-140
Discovery Bay Logging Company 139
Dole Air Race 165
Douglas 154
Douglas, Donald 149
Drain, Col. 80
Dunbar, Marie 156-157
Durant, Cliff 61
Duwamish 9, 41, 46

E

Eads, Jan 162
Ebey's Prairie, Whidbey Island, WA 49
Eckmann, Anscel 113
Edmonds, WA 35, 36
Edwards, Walter 31
Egge, C. F. 112

Egtvedt, Claire 115-116, 145-146, 150-156, 183, 192
Elko, NV 149, 150, 163
Elliott Bay, WA 77
Empress Hotel 59-60, 73, 102, 117, 123, 126, 130
Empress of Asia 136
Empress of Russia 132
Evans, Elwood 136
Everett, WA 43, 65, 114, 185
Evergreen-Washelli 178-179

F

Firby, Charles G. Auctions 39, 199
Fokker Corporation 89-90, 115
Foley, J. C. 26
Ford, Henry 150, 206
Ford Company 150, 154
Ford Trimotor 150
Foreign Air Mail Route No. 2 97, 104, 122, 199
Frink, Mrs. Francis Guy 23, 194
Frink, Gloria 23

G

Galbraith, Jane 115-116
Gardiner, George 102, 123
General Motors 181
Georgetown 32, 90, 150
Gibbs, James 44, 46, 56
Gig Harbor, WA 131
Gonnason, Inez 133, 134
Gorst, Vern 119, 150, 165, 189, 203
Gott, Edgar 28-30, 38, 47, 68-69, 71, 76, 81, 89-90, 95, 110, 145, 149-150, 188, 191
Gott, Mrs. 71
Graham, Mrs. John 178
Graham, Suzanne Carvalho-Maia 37
Grant, Lt. Louis 50-51
Graves, Sandy 57
Great Central Lake, B. C. 136
Green, Mrs. Thomas 194
Green, Senator Robert F. 117-118
Griffin Wheel Co. 181
Gulf Islands, B. C. 134-135

H

Haines, Donald & Loretta 78
Hall-Scott 18, 32, 83, 191-193, 195, 199
Hamilton Metal Plane Division 181
Hamilton Standard Propeller Corp. 181
Hanbury, Evan 133-134
Harrington, Ensign Elliott Dean 56
Hawaiian Clipper 115
Healy, John K. Co. 66-67
Herron, Willard 155-156
Hoffar, Henry 204
Hoffar Shipyards 204
Hoge Building, Seattle 45, 151-152
Hong Kong Clipper 115
Hoy, Capt. 91
Hubbard Air Transport Company 110
Hubbard Field, Reno 166, 168, 189
Hubbard, Margaret 179
Hubbard, Mildred 126, 179
Hubble, Les 47, 110, 193
Huber, Leo 79

I

Inglewood Country Club 78

J

Jackson, Miss Mildred 70
Jacobs, Frank 29, 65
Janes, Harry 80
Jefferson Municipal Golf Course 48, 50, 87-88, 185
Jenny JN-4A Canuck 48-50, 52, 59
Joffre, Marshall Joseph 77, 187
Johnson, Phil 46, 81, 110, 115-116, 149-151, 154-156, 159, 162, 164-165, 167, 181-183, 194, 208
Junkers 114

K

Kashima Maru 63
Kay, Lt. Harry 42
Keena, J. T. 73, 193

Kent, WA 48, 79, 83, 87-88
Kettering, Charles F. 181
King Brothers, Victoria 117
King County, WA 32
Kingsbury, Kenneth R. 181
Knickerbocker, Charles K. 181
Knight, Jack 162,
Knowlton, E. S. 31, 34-35, 121

L

Lake Union, Seattle 9, 25, 28, 32, 35-36, 42, 44-46, 71, 100, 109, 111, 114, 121, 124-128, 130, 132, 140, 183, 193, 200
Lake Washington, Seattle 15-16, 18, 23-24, 46, 71, 77-79, 87, 109, 118, 136, 147, 150, 184, 188, 191
Lane, W. D. 51, 53
Langley, Samuel P. 20
Layton, Mrs. Mildred 46
Leschi Aerial Taxi 76, 97
Liberty Engine 120, 151, 155, 193, 201
Lindbergh, Charles 32, 126, 150, 162, 165-166
Lomen, Alfred 44-46
Lomen, Ralph 44-46
London Daily Mail 129
Los Angeles 150-151, 165, 201-202, 205
Las Vegas 149-150
Luck, Sheldon 204

M

Macpherson, R. Bob 34-35, 38
Madison Park, Seattle 79, 191
Makaru 114
Maroney, Terah 15-16, 19
Marsh, Lou 183, 192
Martin, Glenn 23, 178, 183
Martin M-130 Flying Boat 115
Martin Model TA 23
Maynard Hotel 140
Mayo, William B. 181
McClure, Miss Mildred 133, 135-136
McConachie, Grant 204
McGrath, Edward 95-96, 101
McInnes, Rex 139-140

Mead, George J. 162-163, 181
Meeker Ezra 131-132
Mitchell, General "Billy" 22
Model 40 123-124, 150-153, 155-157, 159, 162, 166-167, 183, 201-202, 204-208
Model 80 124, 149, 164-168, 189, 207-208
Montgomery, John 20
Montreal Maroons 109
Moore, Charles 131
Moore, Townsend Jones 131
Munford, Lt. 31
Munter, Herb 24-26, 29, 48, 79, 83, 87, 95, 114
Museum of Flight, Seattle 5-7, 11, 16, 18, 183, 206, 209
Museum of History & Industry, Seattle 194

N

National Air Transport 182
National City Bank 181
Nelson, Billy 139
New York 115, 136, 151
New York, Rio & Buenos Aires Line 116
New Zealand Airways 204
Newun, Gurney 181
Nicol, Hon. Walter C. 176
North Platte, NB 162-163
Northwest Air Services Inc. 119
Northwestern Air Service 194
Nute, Arthur 24,
Nute, Brud 24, 130

O

Ogden Point, Victoria 142
Olympic Hotel, Seattle 80
Olympic Peninsula, WA 109, 137, 139
Omaha, NB 162, 171
O'Neill, Jane Galbraith 116

P

Pacific Aero Products 16-17, 24, 184
Pacific Air Transport 150, 165, 182, 201-203, 205
Pacific National Bank 179
Pacific Northwest Bell 183

Pan American Airways 115-116
Paramount Pictures 75
Pasco, WA 136, 149-150, 159, 163, 188
Pathfinder 11, 49-50,
Pathfinder Awards 183
Patterson, Mrs. Jimmie 34
Paul, Frank 88, 90
Pavone, Mike 193
Pensacola Bay, FL 146
Perkins, C. M. 34
Philippine Clipper 115
Pierson, E. 61, 70, 193
Pocock, George 191
Poindexter, Senator Miles 43-44, 185
Port Alberni, B. C. 134-135
Port Angeles, WA 80, 140
Port Ludlow, WA 109, 187
Port Madison, WA 137
Port Townsend, WA 49, 63, 70, 185
Porter , Robert 52-53
Praeger, Otto 88, 95
Pratt & Whitney 151-152, 155, 159, 162, 181-182, 201-203, 206
Pratt & Whitney Aircraft Company 181
Priest, Joe 139-140
Puget Sound Airway Company 79
Puyallup, WA 132

R

Race Rocks, B. C. 114
Rae, Juanita 48
Ramsay, Claude 88
Red Barn 209
Red Cross 23, 31, 63, 130, 184-185
Red Crown Gasoline 32-33, 120, 126
Redondo Beach, WA 72
Reno NV 163, 166-169, 189, 212, 214
Reno Cannon International Airport 166, 168
Renton, WA 18, 119, 194, 197
Rentschler, Fred 181
Rich, Silas 113
Rickenbacker, Eddie 12, 61-63, 185
Riddell, Walter 112
Rideout, Lt. Robert 49, 51-53, 59

Ridout, Cec 117
Ripley, Joseph 181
Rithet, J. A. Jack 73, 102, 193
Rithet, R. B. 101, 108, 193
Rithet's Outer Wharf 79, 101-102, 108, 141, 173
Roanoke St., Seattle 44, 76
Rose , Red 204
Ross, Jessica 73, 193
Rotary Club 59-60, 126
Royal Flying Corp. 42, 49, 57
Royal Vancouver Yacht Club 33
Royal Victoria Yacht Club 73
Ryan, Claude 150

S

S. S. President Grant 141-142, 144
S. S. President Jefferson 80-81
S. S. President McKinley 106
S. S. President Madison 78-79
S. S. Prince George 130
Salt Lake City, UT 19, 22, 155-156, 158-159, 162-163, 171, 173, 175-176, 188-189, 194
Salt Lake City Chamber of Commerce 176
Salt Lake City Tribune 19, 22, 172
San Diego Union 27, 150
San Francisco, CA 11, 15-16, 41, 80, 83, 87, 96, 101, 108, 112, 115, 123-124, 149-152, 154-155, 157, 159, 161-166, 182, 184, 186,188, 192, 207
Sand Point, WA 87-90, 124, 136, 150, 153, 156-157, 159, 186-187
Sanford, Mrs. 194
San Juan Island, WA 135
Scott, Clayton 17-18
Seattle 5-7, 9, 11-12, 16, 18, 22-23, 29, 31-32, 34-36, 38, 41-42, 44-46, 49-53, 56-57, 59-61, 63-65, 70, 72-73, 76-81, 84, 87-91, 95-102, 104, 107-117, 119, 120-125, 127-132, 140-144, 149-151, 155-156, 158-159, 164-167, 172-176, 178-179, 182-183, 189, 191, 193-194, 199, 201-203, 205-206, 208-209, 212
Seattle Airplane Construction Company 149
Seattle Chamber of Commerce 38, 89, 150, 155
Seattle Daily Times 22, 42, 55, 57, 72, 77, 89, 96, 137, 158, 172
Seattle Historical Society 194
Seattle Metropolitans 109

Seattle Post-Intelligencer 19, 46, 63, 70, 77, 80-81, 156-157, 172
Seattle Post Office 99, 105, 112, 133
Seattle Transportation Club 87
Seattle-Victoria Air Mail Inc. 119, 212
Sequim,WA 137, 139
Sequim State Bank 139
Severyns, Bill 79-80
Shanghai 79-80, 99
Shawnigan Lake, B. C. 131
Shriners 80
Sidney, B.C. 114
Sikorsky, S-42B Flying Boat 115
Sikorsky Aviation Corporation 182
Smith, Capt. George 138
Smith, Gerald 12, 106, 117-118, 127
Smith, Mara 183
Smith, Miss Vera 44, 46,
Society of Automotive Engineers 178
Sperry, Lawrence 21
St. Michaels University School 130
St. Joseph's Hospital 117
Standard Furniture Co. 158
Standard J Aircraft 125
Standard Oil Company of California 32, 36, 124, 126, 203, 208
Stanley Cup 109
Stanley Fish Reduction Co. 109
Starwich, Matt 140
Stearman Aircraft Company 182
Stout Air Services, Inc. 154, 182
Stout Metal Airplane Company 150
Strain, William 119
Straits of Juan de Fuca 114
Strang, F. W. 64, 70
Strang & Prosser 64, 193
Stanwood 68-70
Stromer, Gustave 31
Sunbeam Coal Mines 119
Suwa Maru 70

T

Tacoma, WA 31, 46, 56, 61, 71, 114, 185-186
Taconite 61, 73-74
Tetrazzini, Madame 72, 186

The Daily Colonist 55, 77, 172
The Seattle Star 139
Tupper, Orville 155-156, 181

U

Umbrecht, Mrs. Betty 23
United Air Lines 168, 181-182, 202
United Air Transport Ltd. 204
United Aircraft and Transportation Corporation 181
United States 193-195
United States Army 49-50
United States Coast Guard 139
United States Foreign Air Mail 122, 199
United States Navy 33, 121, 199
United States Post Office 112, 119, 144, 167
Universal News Service 80
University of Washington 7, 70, 77, 194
U. S. West Communications 182-183

V

Van Welch, Chief 140
Vancouver, B. C. 32-33, 36, 41, 91, 98, 114, 121, 126, 133, 151, 199, 204
Vancouver Board of Trade 37
Vancouver Daily Sun 35
Vancouver Island 126
Van Devoort, W. C. 113
Varney, Walter 150
Varney Air Lines 163, 182
Vickers Vimy 129
Victoria, B. C. 9, 23, 35, 47, 49-50, 52-53, 56-57, 59-60, 63, 70, 73, 77-81, 8-89, 95-102, 104, 107-115, 119-120, 122-123, 125, 127, 130-134, 136, 141, 144, 151, 176, 182, 193, 199
Victoria Colonist 77, 172, 176
Victoria Daily Times 57, 71, 81, 97, 111, 113, 117, 141, 171-172, 172, 175, 186
Victoria Festival Society 52
Victoria Gas Company 77
Victoria Machinery Depot 143
Victoria Outer Wharf 79
Victoria Post Office 99, 105
Vought 181

W

Wadsworth, E. B. 162
Waterhouse, Norman 138
Weiler, George 108, 133-134, 142
Weiler, Mary 108
Weiler, Sis 133-135
Weller, Otto Jr. 108
Wells, Linton 136
Western Air Express 154, 163
Westervelt, G. C. 16
Whidbey Island, WA 5, 12, 49-50, 122
Williams, George 113, 133
Willows Fairgrounds 52
Willows Hotel, Campbell River 74
Wright Aeronautical Corp. 177
Wright Brothers 20
Wygle, Brien 10

About the Author

Jim Brown's interest in Eddie Hubbard sprang from his research on air mail and his avid interest in stamp collecting. He has written articles on air mail flights for several magazines and newspapers and has a regular column of aircraft on stamps in the *West Coast Aviator.*

In his youth, he served in the Royal Canadian Navy. With such a great interest in airplanes, he was often asked why he didn't join the Air Force. At the time, he explains, all they were recruiting were tail gunners and for some reason that held absolutely no appeal.

After his discharge, he attended the University of British Columbia. His career choice was in marketing and he was with Standard Oil of B.C. and Chevron Oil Europe headquarters in Brussels until 1984.

He is an avid golfer, an ardent jazz and big band buff, and a life long fisherman. He and his wife Gwen live on Pender Island, B. C. and enjoy frequent visits from their six children and nine, soon-to-be eleven, grandchildren.

Order Form

To order additional copies of:

Hubbard: The Forgotten Boeing Aviator

Please send $19.95 or $24.95 in Canada
plus $3.00 Shipping & Handling.
(Canadian residents add 7% GST.
Washington residents please include 8.2% sales tax.)
Make check or money order payable to:
Peanut Butter Publishing
226 2nd Ave W.
Seattle, WA 98119
(206) 281-5965

If you prefer to use VISA or Mastercard, please fill in your card's number and expiration date. Please circle the appropriate card.

☐☐☐☐☐☐☐☐☐☐☐☐☐☐☐☐

Signature____

exp. date_______

______Copies @ $19.95ea.________
Canada $24.95ea.________
$3.00 Shipping & Handling________
Washington State residents add 8.2%_______
Canadian residents add 7% GST _______
Total enclosed_______

Name____
Address______________________________________
City, State, Zip_________________________________

Please list additional copies to be sent to other addresses on a separate sheet.